Holt Mathematics

Chapter 9 Resource Book

HOLT, RINEHART AND WINSTON
A Harcourt Education Company
Orlando • Austin • New York • San Diego • London

Copyright © by Holt, Rinehart and Winston

All rights reserved. No part of this publication may be reproduced or transmitted in any form or by any means, electronic or mechanical, including photocopy, recording, or any information storage and retrieval system, without permission in writing from the publisher.

Teachers using HOLT MATHEMATICS may photocopy complete pages in sufficient quantities for classroom use only and not for resale.

Printed in the United States of America

If you have received these materials as examination copies free of charge, Holt, Rinehart and Winston retains title to the materials and they may not be resold. Resale of examination copies is strictly prohibited and is illegal.

Possession of this publication in print format does not entitle users to convert this publication, or any portion of it, into electronic format.

ISBN 0-03-078399-2

CONTENTS

Blackline Masters

Parent Letter	1
Lesson 9-1 Practice A, B, C	3
Lesson 9-1 Reteach	6
Lesson 9-1 Challenge	7
Lesson 9-1 Problem Solving	8
Lesson 9-1 Reading Strategies	9
Lesson 9-1 Puzzles, Twisters & Teasers	10
Lesson 9-2 Practice A, B, C	11
Lesson 9-2 Reteach	14
Lesson 9-2 Challenge	16
Lesson 9-2 Problem Solving	17
Lesson 9-2 Reading Strategies	18
Lesson 9-2 Puzzles, Twisters & Teasers	19
Lesson 9-3 Practice A, B, C	20
Lesson 9-3 Reteach	23
Lesson 9-3 Challenge	24
Lesson 9-3 Problem Solving	25
Lesson 9-3 Reading Strategies	26
Lesson 9-3 Puzzles, Twisters & Teasers	27
Lesson 9-4 Practice A, B, C	28
Lesson 9-4 Reteach	31
Lesson 9-4 Challenge	32
Lesson 9-4 Problem Solving	33
Lesson 9-4 Reading Strategies	34
Lesson 9-4 Puzzles, Twisters, & Teasers	35
Lesson 9-5 Practice A, B, C	36
Lesson 9-5 Reteach	39
Lesson 9-5 Challenge	41
Lesson 9-5 Problem Solving	42
Lesson 9-5 Reading Strategies	43
Lesson 9-5 Puzzles, Twisters & Teasers	44
Lesson 9-6 Practice A, B, C	45
Lesson 9-6 Reteach	48
Lesson 9-6 Challenge	49
Lesson 9-6 Problem Solving	50
Lesson 9-6 Reading Strategies	51
Lesson 9-6 Puzzles, Twisters & Teasers	52
Lesson 9-7 Practice A, B, C	53
Lesson 9-7 Reteach	56
Lesson 9-7 Challenge	57
Lesson 9-7 Problem Solving	58
Lesson 9-7 Reading Strategies	59
Lesson 9-7 Puzzles, Twisters & Teasers	60
Lesson 9-8 Practice A, B, C	61
Lesson 9-8 Reteach	64
Lesson 9-8 Challenge	65
Lesson 9-8 Problem Solving	66
Lesson 9-8 Reading Strategies	67
Lesson 9-8 Puzzles, Twisters & Teasers	68
Answers to Blackline Masters	69

Holt Mathematics

Date_____

Dear Family,

In this chapter, your child will learn to collect, display, and analyze data. These are critical skills today when information technology, such as the computer and the Internet, has become a vital part of the nation's economy.

Collecting data scientifically is important in order to get accurate results. A population is the entire group being studied. A sample is the part of the population being surveyed. The **sampling method** used to collect data needs to produce a sample that represents the population.

Sampling Method	How Sample Members Are Chosen
Random	By chance
Systematic	According to a rule or formula
Stratified	At random from randomly chosen subgroups

Organizing data is the next step. Once you collect accurate data, you need to organize the data so others can read and interpret it. A table is a useful tool for organizing data.

	Job 1	Job 2	Job 3
Salary Range	$20,000–$34,000	$18,000–$50,000	$14,000–$40,000
Benefits	$12,000	$10,500	$11,400
Vacation Days	10	15	12

There are many types of **graphs** for **displaying data**. Your child will learn to create histograms such as the one below, which shows the number of books of varying lengths that a class read.

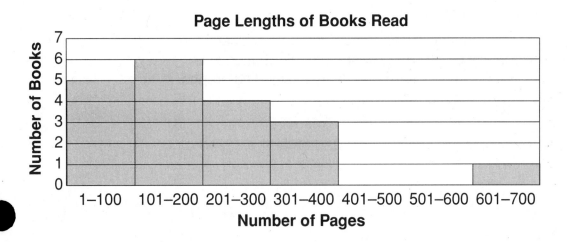

Holt Mathematics

Finally, understanding **mean, median,** and **mode** is important to analyzing data. These statistics are used to describe the "center" of a data set.

Measures of Central Tendency

	Definition	Use to answer
Mean	The sum of the values, divided by the number of values	"What is the average?"
Median	If an odd number of values, the middle value. If an even number of values, the average of the two middle values	"What is the halfway point of the data?"
Mode	The value or values that occur most often	"What is the most frequent value?"

Find the mean, median, and mode of the following data set.
4, 8, 8, 3, 6, 8, 3

The **mean** is the sum of the values, divided by the number of values.

mean: $4 + 8 + 8 + 3 + 6 + 8 + 3 = 40$ Add the values.

$\frac{40}{7} \approx 5.7$ Divide by 7, the number of values.

The **median** is the halfway point in the data, or the middle value.

median: 3 3 4 6 8 8 8

    three values   three values

The median is 6.

The **mode** is the value that occurs most often.

mode: 8   8 occurs three times.

Graphs and statistics are often used to persuade. Advertisers and others may accidentally or intentionally present information in a misleading way. Your child will learn to identify misleading graphs and to explain how the data is inaccurate.

For additional resources, visit go.hrw.com and enter the keyword MT7 Parent.

Name _____ Date _____ Class _____

LESSON 9-1 Practice A
Samples and Surveys

Identify the sampling method used.

1. The name of each audience member at a game show is printed on a separate piece of paper and put into a rotating case. One audience member is chosen to play the game by drawing one piece of paper from the shuffled names.

   _____

2. At a frozen pizza manufacturing plant, a coupon for a free pizza is put inside the package of every 100th pizza.

   _____

3. The teacher asks that students with birthdays from July to December go to the chalkboard to work the next problem.

   _____

Identify the population and sample. Give a reason the sample could be biased.

4. A radio disc jockey asks listeners to call in and name their favorite radio station.

   population _____

   sample _____

   possible bias _____

5. A hospital mails out surveys to 500 recent patients to get their feedback on their hospital visit.

   population _____

   sample _____

   possible bias _____

6. The first 10 people leaving the theater are asked to give their feedback about the movie.

   population _____

   sample _____

   possible bias _____

Name _____ Date _____ Class _____

LESSON 9-1 Practice B
Samples and Surveys

Identify the sampling method used.

1. People in the security line at the airport are asked to step out of the line for a more detailed search. The people pulled out of the line have not necessarily done anything wrong, and they are not chosen according to any particular rule.

   _____

2. At the 1-mile marker of a marathon, a timekeeper shouts out the time elapsed to every 10th runner that passes by. A statistician records the times shouted.

   _____

3. A geologist visits 10 randomly-selected lakes in the region and collects soil samples in randomly-selected areas along each shoreline.

   _____

Identify the population and sample. Give a reason the sample could be biased.

4. At a convention of science teachers, various attendees are asked to name their favorite subject in high school.

   population _____

   sample _____

   possible bias _____

5. Donors participating in a blood drive are given a small amount of money for their blood donation. Before they can give blood, each person is surveyed to find out if they are eligible to give blood.

   population _____

   sample _____

   possible bias _____

6. Interviewers at the mall are surveying girls with red hair to find out if a correlation exists between personality and red hair.

   population _____

   sample _____

   possible bias _____

Name _____ Date _____ Class _____

LESSON 9-1 Practice C
Samples and Surveys

Identify the sampling method used.

1. In clinical research done on a new medication, every other patient is given a placebo. (A placebo is an inactive substance that does not have the effect of the medicine. Placebos are used in controlled medical experiments.)

   _____

2. A television station sends out reporters to randomly-selected neighborhoods throughout the city. In each neighborhood randomly-selected voters are polled to find the approval rating of the current mayor.

   _____

3. A surveyor calls the last person on each page of a phone book.

   _____

Identify the population and sample. Give a reason the sample could be biased.

4. The professional baseball league All-Star ballots are handed out at the stadium on game days.

   population _____

   sample _____

   possible bias _____

5. The city mayor visits a neighborhood council meeting and asks for a raise of hands to show support for his education agenda.

   population _____

   sample _____

   possible bias _____

6. An online newspaper asks every person who reads a certain article to rate the article.

   population _____

   sample _____

   possible bias _____

Name _____ Date _____ Class _____

LESSON 9-1
Reteach
Samples and Surveys

A *survey* uses a *small sample* to represent a *large population*.

Sampling Method	How Members of the Sample Are Chosen
Random	By chance; members have an equal chance of selection
Systematic	According to a rule or pattern
Stratified	At random from randomly-chosen subgroups

A senator's office sends workers to ask constituents at a local mall how they feel about floating a bond to acquire land around a reservoir to be preserved as open space.

The population in this survey are all the eligible voters of the state. The shoppers at the mall are the sample of that population.

Identify the sampling method used.

1. A survey calls every 10th name listed in a local phone book.

   _____

Identify the population and the sample.

2. Kennedy HS seniors planning to attend the prom were asked if senior dues should include a photo taken at the prom.

   Population _____

   Sample _____

A *biased sample* is not a good representative of the population.

A school principal asks parents attending an art workshop if funding for a theater arts program should be included in the school budget.

This is a biased sample since the parents attending an art workshop are likely to be in favor of additional art programs.

Identify the population and the sample. Give a reason why the sample could be biased.

3. Homeowners within a 10-mile radius of a nuclear power plant were asked if they think the plant should be closed.

   Population _____

   Sample _____

   Possible Bias: _____

   _____

Copyright © by Holt, Rinehart and Winston.
All rights reserved.

Holt Mathematics

Challenge

Lesson 9-1: Stay in the Margins

A survey gathers information from a few people and then the results are used to reflect the opinions of a larger population. Researchers and pollsters use sample populations because it is cheaper and easier to poll a few people than to ask everybody. One key to successful surveys of sample populations is finding the appropriate size for the sample that will give accurate results without spending too much time or money.

Suppose that 900 American teens were surveyed about their favorite ski category of the 2002 Winter Olympics in Park City, Utah. Ski Jumping was the favorite for 20% of those surveyed. This result can be used to predict how many of all 31 million American teens favor Ski Jumping.

$$31{,}000{,}000 \times 0.20 = 6{,}200{,}000 \text{ American teens favor Ski Jumping}$$

To determine how accurately the results of surveying 900 American teens truly reflect the results of surveying all 31 million American teens, a **margin of error** should be given. When pollsters report the margin of error for their surveys, they are stating their confidence in the data they have collected.

The margin of error can be calculated by using the formula $\frac{1}{\sqrt{n}}$, where n is the number in a sample size.

For the above sample the margin of error would be $\frac{1}{\sqrt{900}} = \frac{1}{30} = 0.03\overline{3} \approx 3\%$. Since the actual statistic could be smaller or larger than the true amount, the margin of error is expressed as ±3%.

1. Find the margin of error for a survey of 100 American teens. _____

2. Compare that margin of error to the margin of error of 900 teens.

   _____

3. Find the margin of error for a survey of 9,000 American teens. _____

4. Find the margin of error for a survey of 90,000 American teens. _____

5. Draw a conclusion about the margin of error based on size of the sample.

LESSON 9-1: Problem Solving
Samples and Surveys

Identify the sampling method used.

1. Every twentieth student on a list is chosen to participate in a poll.

2. Seat numbers are drawn from a hat to identify passengers on an airplane that will be surveyed.

Give a reason why the sample could be biased.

3. A company wants to find out how its customers rate their products. They ask people who visit the company's Web Site to rate their products.

4. A teacher polls all of the students who are in detention on Friday about their opinions on the amount of homework students should have each night.

A car dealership wants to know how people who have visited the dealership feel about the dealership and the sales people. They survey every 5th person who buys a car. Choose the letter for the best answer.

5. Identify the population.
   A People who visit the dealership
   B People who buy a car from the dealership
   C People in the local area
   D The salesmen at the dealership

6. Identify the sample.
   F Every person who visits the dealership
   G People who buy a car
   H Every 5th buyer
   J People in the local area

7. Identify the possible bias.
   A Not all people will visit the dealership.
   B Did not survey everyone who buys a car.
   C Not including those who visited but did not buy.
   D There is no bias.

8. Identify the sampling method used.
   F Random
   G Systematic
   H Stratified
   J None of these

Name _____ Date _____ Class _____

LESSON 9-1 Reading Strategies
Compare and Contrast

Surveys are taken to get information about a group of people. A survey may include:
- the entire group, called the **population**;
- part of the group, called a **sample** of the population.

Compare a population to a sample.

1. How are a sample and a population alike?

   _____

   _____

2. How are a population and a sample different?

   _____

   _____

Sometimes it is impossible to survey an entire population. Most of the time only a sample of the population is surveyed. There are two kinds of samples.
- An **unbiased sample** accurately represents the population.
- A **biased sample** does not represent the entire population.

A company wants to test market a new sports drink for high school students. Write "unbiased sample" or "biased sample" for each situation.

3. If the company surveys every tenth student in seven different high schools across the country, it would be a(n) _____.

4. If the company surveys only athletes at the seven high schools, it would be a(n) _____.

Compare an unbiased sample to a biased sample.

5. How is an unbiased sample like a biased sample?

   _____

   _____

6. How is an unbiased sample different from a biased sample?

   _____

   _____

Name _____ Date _____ Class _____

LESSON 9-1 Puzzles, Twisters & Teasers
Meltdown!

Find and circle the words below in the word search (horizontally, vertically or diagonally). Find a word that answers the puzzle. Circle the word and write it on the line.

population	sample	biased	random	systematic
stratified	method	average	exit	poll

```
S Y S T E M A T I C P O L L
T R A B T E V U I B U G B R
R B M I R T E P O I D D C A
A M P A E H R X S W D U H N
T I L S W O A M N B L N J D
I O E E Q D G P L O E I O O
F W S D F G E C D E A K L M
I Q W E R T Y U I O X A S D
E P O T C V B N M X Z I Q W
D P O P U L A T I O N T T E
```

What do you call a snowman with a tan?

A _____ .

Name _____ Date _____ Class _____

Practice A
LESSON 9-2 Organizing Data

1. Complete the line plot to organize the data of math quiz scores.

Math Quiz Scores

18 18 20 13 17 12 15 12
17 19 17 18 18 20 11 19

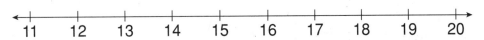

List the data values in the stem-and-leaf plot.

2. 0 | 1 2 5
   1 | 0 5
   2 | 2 4 6
   3 | 1 7   Key: 3 | 7 = 37

   _____
   _____
   _____

3. Use the given data to make a stem-and-leaf plot.

Maximum Speed of Animals (mph)			
pig (domestic)	11	grizzly bear	30
squirrel	12	rabbit	35
elephant	25	zebra	40
cat (domestic)	30	cape hunting dog	45

Key:

4. Make a Venn diagram to show how many boys in an eighth-grade class had summer jobs.

Gender	M	M	F	M	F	F	M	F	F	M	M	M
Summer Job?	yes	no	yes	yes	yes	no	no	yes	yes	yes	no	yes

Holt Mathematics

Practice B
9-2 Organizing Data

1. Use a line plot to organize the data of the distances students travel to school.

   Distances Students Travel to School (mi)
2	8	6	10	5	4	6	8	3	2
11	5	1	3	6	5	7	5	2	4

List the data values in the stem-and-leaf plot.

2. 2 | 0 1 5 7
   3 | 2 2 9
   4 | 5 6 7 9
   5 | 1 3   Key: 5 | 1 = 51

   _____

3. Use the given data to make a back-to-back stem-and-leaf plot.

NBA Midwest Division 2000–2001 Final Standings

NBA Team	Wins	Losses	NBA Team	Wins	Losses
San Antonio Spurs	58	24	Houston Rockets	45	37
Utah Jazz	53	29	Denver Nuggets	40	42
Dallas Mavericks	53	29	Vancouver Grizzlies	23	59
Minnesota Timberwolves	47	35			

Wins | Losses

Key:

4. Make a Venn diagram to show how many girls in an eighth-grade class belonged to both a team and a club.

Team	yes	no	yes	no	yes	yes	yes	no	no	yes	no	no
Club	yes	yes	no	yes	yes	no	yes	yes	yes	no	no	yes

Name _____ Date _____ Class _____

LESSON 9-2 Practice C
Organizing Data

1. Use a line plot to organize the data of dog temperatures measured at a veterinary's office one day.

   Dog Temperatures (F°)

   | 101.3 | 101.6 | 101.5 | 102.3 | 101.9 | 102.1 |
   | 101.5 | 102.2 | 102.0 | 101.8 | 102.1 | 101.8 |

List the data values in the stem-and-leaf plot.

2.
```
0 | 6
1 | 1 5 6 8
2 | 3
3 | 0 7 9
4 | 1 2 4      Key: 4 | 1 = 4.1
```

3. Use the given data to make a back-to-back stem-and-leaf plot.

Animal	Endangered	Threatened	Endangered		Threatened
Mammals	63	9			
Birds	78	14			
Reptiles	14	22			
Amphibians	10	8			
Fishes	70	44			
Clams	61	8			
Snails	20	11			
Insects	33	9	Key:		
Arachnids	12	0			

4. Make a Venn diagram to show how many students in an eighth-grade class have both a television and a computer in their bedroom.

Television	yes	yes	no	no	no	yes	yes	yes	yes	yes	no	yes
Computer	yes	yes	no	yes	no	yes	no	yes	yes	no	yes	no

Name _____ Date _____ Class _____

LESSON 9-2

Reteach
Organizing Data

Horizontal displays of numbers can be written in a compact form by eliminating repetition.

60 61 63 66 66 67 can be written 6 | 0 1 3 6 6 7

All the numbers start with 6, the **stem**.

Just write the second digit for each number, the **leaves**.

Display each set of numbers in compact form, using a stem and leaves.

1. 72 75 75 76 76 76 79

2. 120 123 124 125 125 127

_____   _____

Here are the scores on the last test in Ms. Kahn's math class.
76 84 88 93 97 65 100 86 91 97
93 79 81 99 92 78 78 79 87 100

To display these scores in a **stem-and-leaf plot**:

Use the given order to record scores.

```
 6 | 5
 7 | 6 9 8 8 9
 8 | 4 8 6 1 7
 9 | 3 7 1 7 3 9 2
10 | 0 0
```
Key: 7 | 6 represents 76.

Order each set of leaves.

```
 6 | 5
 7 | 6 8 8 9 9
 8 | 1 4 6 7 8
 9 | 1 2 3 3 7 7 9
10 | 0 0
```
Key: 7 | 6 represents 76.

Complete a stem-and-leaf plot for the data.

3. Daily High Temperatures

46 52 48 47 56 59 61 50 37 35 34 37 44 49 43
43 44 50 50 52 53 50 48 46 44 39 37 32 32 32

```
3 | 7 5 4 7 9 7 2 2 2
4 |
5 |
6 |
```

```
3 | 2 2 2 4 5 7 7 7 9
4 |
5 |
6 |
```

Key: _____ represents 52.   Key: _____ represents 52.

4. Create a stem-and-leaf plot for heights in inches of students in Mrs. Gray's class.
40 48 49 53 60 62 48 62 55 53 54 60 65 63 49 55

Name _____ Date _____ Class _____

LESSON 9-2 Reteach
Organizing Data (continued)

Two sets of data can be compared by using a **back-to-back stem-and-leaf plot**.

This plot compares the number of games won to the number of games lost by each of the pennant winners in the American League East for the years 1995–2000.

The stems are in the center.
The left leaves are read in reverse.

The greatest number of games won was 114.

The greatest number of games lost was 74.

**American League East
Pennant Winners 1994–2000**

Games Lost		Games Won
8	4	
8	5	
4 4	6	
4 0	7	
	8	6 7
	9	2 8 8
	10	
	11	4

Key: | 8 | 6   represents 86 games.
Key: 8 | 4 | represents 48 games.

Refer to the American League East back-to-back stem-and-leaf plot shown above.

5. The least number of games won was: _____

6. The least number of games lost was: _____

Using the given data set, complete to make a back-to-back stem-and-leaf plot.

7.   **American League West
Pennant Winners**

Year	Winner	Won	Lost
1994	Texas	52	62
1995	Seattle	79	66
1996	Texas	90	72
1997	Seattle	90	72
1998	Texas	88	74
1999	Texas	95	67
2000	Oakland	91	70

Games Lost		Games Won
	5	
	6	
	7	
	8	
	9	

Key: represents 88 games.
Key: represents 62 games.

Refer to the American League West back-to-back stem-and-leaf plot.

8. The difference between the greatest number of games won and the least number of games won was: _____

9. The difference between the greatest number of games lost and the least number of games lost was: _____

Name _____ Date _____ Class _____

Challenge
LESSON 9-2 Pet Sets

A Venn diagram can show the relationships among three sets of data. Use the survey results shown in the table to complete the Venn diagram.

Number of Students in Mr. Phillips' Math Class Whose Families Have Pets						
Dogs	Dogs and Cats	Dogs and Fish	Dogs, Cats, and Fish	Cats	Cats and Fish	Fish
13	7	2	2	12	3	6

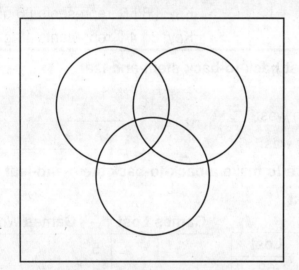

Use your Venn diagram to answer the questions.

1. How many of the families have dogs only? _____

2. How many of the families have cats only? _____

3. How many of the families have fish only? _____

4. How many of the families have dogs and fish but no cats? _____

5. How many of the families have cats and fish but no dogs? _____

Name _____ Date _____ Class _____

LESSON 9-2 Problem Solving
Organizing Data

A consumer survey gathered the following data about what teens do while on online.

1. Make a stem-and-leaf plot of the data.

Teens' Activities Online	
Activity	Percent
E-mail	95
Use search engines	86
Instant Messaging	82
Visit music sites	73
Enter contests	73

The stem-and-leaf plot that shows the total number of medals won by different countries in the 2000 Summer Olympics. Choose the letter for the best answer.

2. List all the data values in the stem-and-leaf plot.
   A 2, 4, 5, 6, 7, 8, 9
   B 23, 25, 26, 28, 28, 29, 34, 38, 40, 57, 58, 59, 60, 70, 88, 97
   C 23, 25, 26, 28, 29, 34, 38, 57, 58, 59, 88, 97
   D 23, 25, 26, 28, 28, 29, 34, 38, 57, 58, 59, 88, 97

2000 Olympic Medals

```
2 | 3 5 6 8 8 9
3 | 4 8
4 |
5 | 7 8 9
6 |
7 |
8 | 8
9 | 7
```

3. What is the least number of medals won by a country represented in the stem-and-leaf plot?
   F 3
   G 4
   H 23
   J 97

4. What is the greatest number of medals won by a country represented in the stem-and-leaf plot?
   A 9
   B 70
   C 79
   D 97

Name _____ Date _____ Class _____

LESSON 9-2 Reading Strategies
Use a Graphic Organizer

Below is a list of students' scores for a science test.

75 84 90 68 73 83 95 85 89 97 77 72 83

A stem-and-leaf plot is one way to organize the test scores.

stem	leaf
6	5 8
7	2 3 5 7
8	3 3 4 5 9
9	0 5 7

Use your stem-and-leaf plot to answer the questions.

1. What is the stem of the score 83? _____
2. What is the leaf of the score 90? _____
3. What is the highest score? _____
4. What is the lowest score? _____
5. What is the difference between the highest and lowest scores? _____

Below is a list of students' scores for a social studies test.

69 84 75 99 65 70 81 87 73 81 92 95 71

6. Make a stem-and-leaf plot of the scores.

stem	leaf
6	
	0

7. Can you find the lowest and highest scores more easily by looking at the list or at the stem-and-leaf plot?

_____

Puzzles, Twisters & Teasers
9-2 A Bright Idea!

List the values in the stem-and-leaf plot from the top line down. Each answer has a corresponding letter. Use the letters to solve the riddle.

```
3 | 2 5
4 | 0 3 7
5 | 1 4 8
6 | 1 6
```

T ___32___
H ___35___
W ___40___
E ___43___
G ___47___
R ___51___
I ___54___
S ___58___
B ___61___
O ___66___

Why did the teacher wear sunglasses in class?

Because her students __W__ __E__ __R__ __E__   __S__ __O__   __B__ __R__ __I__ __G__ __H__ __T__.
                          40 43 51 43   58 66   61 51 54 47 35 32

Name _____ Date _____ Class _____

Practice A
LESSON 9-3 *Measures of Central Tendency*

Find the mean, median, mode, and range of each set of numbers.

1. 4, 2, 6, 3, 8, 6, 6

    mean: _____   mode: _____

    median: _____   range: _____

2. 2, 8, 6, 9, 8, 7, 9, 8

    mean: _____   mode: _____

    median: _____   range: _____

3. 12, 9, 14, 22, 3, 11, 14, 15

    mean: _____   mode: _____

    median: _____   range: _____

4. 89, 45, 68, 94, 70, 94, 86

    mean: _____   mode: _____

    median: _____   range: _____

Determine and find the most appropriate measure of central tendency or range for each situation. Refer to the table.

5. What number best describes the middle of the waterfall heights?

    _____

6. What number appears most often in the waterfall heights?

    _____

7. Which measure of central tendency is best to describe the waterfall heights? Explain your reasoning.

    _____

    _____

Waterfall Heights (ft)	
Feather, CA	640
Bridalveil, CA	620
Ribbon, NV	1,612
Seven, CO	300
Akaka, HI	442
Shoshone, ID	212
Taughannock, NY	215
Multnomah, OR	620

8. An official for the department of transportation counted the number of vehicles that passed through a busy intersection. He counted for 10 consecutive minutes and recorded the number of vehicles for each minute: 18, 41, 25, 9, 22, 36, 24, 13, 25, and 28. What number best describes the middle of the data?

    _____

Name _____ Date _____ Class _____

Practice B
9-3 Measures of Central Tendency

Find the mean, median, mode, and range of each data set.

1. 7, 7, 4, 9, 6, 4, 5, 8, 4

   mean: _____

   median: _____

   mode: _____

   range: _____

2. 1.2, 5.8, 3.7, 9.7, 5.5, 0.3, 8.1

   mean: _____

   median: _____

   mode: _____

   range: _____

3. 31, 28, 31, 30, 31, 30,
   31, 31, 30, 31, 30, 31

   mean: _____

   median: _____

   mode: _____

   range: _____

4. 65, 46, 78, 3, 87,
   12, 99, 38, 71, 38

   mean: _____

   median: _____

   mode: _____

   range: _____

Determine and find the most appropriate measure of central tendency or range for each situation. Refer to the table at the right for Exercises 5–7.

Some Major Earthquakes in United States History

Year	Location	Magnitude
1812	Missouri	7.9
1872	California	7.8
1906	California	7.7
1957	Alaska	8.8
1964	Alaska	9.2
1965	Alaska	8.7
1983	Idaho	7.3
1986	Alaska	8.0
1987	Alaska	7.9
1992	California	7.6

5. Which measure best describes the middle of the data?

   _____

6. Which earthquake magnitude occurred most frequently?

   _____

7. How spread out are the data?

   _____

8. Nicole purchased gasoline 8 times in the last two months. The prices that she paid per gallon each time were $2.19, $2.14, $2.28, $2.09, $2.01, $1.99, $2.19, and $2.39. Which measure makes the prices appear lowest?

   _____

Name _____ Date _____ Class _____

LESSON 9-3 Practice C
Measures of Central Tendency

Find the mean, median, mode, and range of each data set.

1. 94, 90, 88, 66, 94, 81, 102, 108, 88

   mean: _____

   median: _____

   mode: _____

   range: _____

2. 16.58, 9.99, 24.30, 48.85, 25.09, 9.71

   mean: _____

   median: _____

   mode: _____

   range: _____

3. 173, 160, 232, 148, 162, 160, 265, 182, 98, 147, 162, 205, 162, 169

   mean: _____

   median: _____

   mode: _____

   range: _____

4. 2386, 3154, 2873, 4256, 3184, 2389, 3141, 2452, 3000, 2584, 4189

   mean: _____

   median: _____

   mode: _____

   range: _____

Determine and find the most appropriate measure of central tendency or range for each situation. For Exercises 5–7, refer to the table at the right.

5. Which measure best describes the middle of the data?

   _____

6. Which difference in altitude occurs most frequently?

   _____

7. How spread out are the data?

   _____

Difference Between the Highest and Lowest Points of Each Continent

Continent	Difference (Feet)
North America	20,602
South America	22,965
Europe	18,602
Asia	30,347
Africa	19,852
Australia & Oceania	7,362
Antarctica	25,191

8. As of 2001, the estimated population, in millions, of each continent is as follows: North America, 476; South America, 343; Europe, 727; Asia, 3641; Africa: 778; Australia & Oceania, 30; and Antarctica, 0. Which measure makes the population appear greatest?

   _____

Name _____ Date _____ Class _____

LESSON 9-3 Reteach
Measures of Central Tendency

The **mode** of a data set is the value (or values) that occur(s) most often.

　　　2, 4, 10, **3**, 6, **3**, 7
　　　The value 3 occurs most often. So, 3 is the mode.

The **median** of a data set is the middle value—after the values have been ordered.

　　　2, 4, 10, 3, 6, 3, 7　⟶　2, 3, 3, **4**, 6, 7, 10
　　　The middle value is 4. So, 4 is the median.

The **mean** of a data set is the average value. Add the values and divide the sum by the number of values in the set.

　　　2, 4, 10, 3, 6, 3, 7　⟶　$\frac{2 + 4 + 10 + 3 + 6 + 3 + 7}{7} = \frac{35}{7}$, or 5
　　　So, the mean is 5.

The **range** of a data set is the difference between the greatest value and the least value.

　　　2, 4, **10**, 3, 6, 3, 7　⟶　10 − 2 = 8
　　　So, the range is 8.

Determine and find the most appropriate measure of central tendency or range for each situation. Refer to the table at the right.

1. Which age was most frequent at the reunion?

　　_____

2. Which age was the average at the reunion?

　　_____

3. How spread out are the ages?

　　_____

4. What was the middle age?

　　_____

Smith Family Reunion 2006

Family Member	Age
Aunt Beth	36
Uncle Steve	40
Louise	9
Travis	6
Grandma	62
Grandpa	62
Mom	44
Dad	43
Me	13

Name _____ Date _____ Class _____

LESSON 9-3 Challenge

The Groupie Effect

A class of 29 students reported the number of books read so far this school year.

Number of Books Read
5, 5, 6, 3, 6, 3, 2, 7, 5, 3,
7, 4, 2, 5, 6, 7, 6, 4, 1, 4,
9, 5, 6, 7, 7, 6, 6, 7, 5

This frequency table shows the same data.

Value	1	2	3	4	5	6	7	8	9
Frequency	1	2	3	3	6	7	6	0	1

1. Explain how to find the mode using the ungrouped data. What is the mode?

   _____

2. Explain how to find the mode using the frequency table. Verify that you obtain the same result as before.

   _____

3. Explain how to find the median using the ungrouped data. What is the median?

   _____
   _____

4. Explain how to find the median using the frequency table. Verify that you obtain the same result as before. Which method do you prefer? Why?

   _____
   _____

5. Explain how to find the mean using the ungrouped data. What is the mean?

   _____

6. Explain how to find the mean using the frequency table. Verify that you obtain the same result as before. Which method do you prefer? Why?

   _____
   _____

Name _____ Date _____ Class _____

Problem Solving
LESSON 9-3 Measures of Central Tendency

Use the data to find each answer.

World's Busiest Airports

Airport	Total Passengers (in millions)
Atlanta, Hartsfield	80.2
Chicago, O'Hare	72.1
Los Angeles	68.5
London, Heathrow	64.6
Dallas/Ft. Worth	60.7

1. Find the average number of passengers in the world's five busiest airports.

   _____

2. Find the median number of passengers in the world's five busiest airports.

   _____

3. Find the mode of the airport data.

   _____

4. Find the range of the airport data.

   _____

Choose the letter for the best answer.

World Motor Vehicle Production (in thousands) 1998–1999

Country	1998	1999
United States	12,047	13,063
Canada	2,568	3,026
Europe	16,332	16,546
Japan	10,050	9904

5. What was the mean production of motor vehicles in 1998?
   A 8,651,500 vehicles
   B 10,249,250 vehicles
   C 11,264,250 vehicles
   D 12,000,000 vehicles

6. What was the range of production in 1999?
   F 9,800,000 vehicles
   G 11,480,000 vehicles
   H 12,520,000 vehicles
   J 13,520,000 vehicles

7. What was the median number of vehicles produced in 1999?
   A 3,026,000 vehicles
   B 3,069,000 vehicles
   C 11,483,500 vehicles
   D 13,063,000 vehicles

8. Which value is largest?
   F Mean of 1998 data
   G Mean of 1999 data
   H Median of 1998 data
   J Median of 1999 data

Holt Mathematics

Name _____ Date _____ Class _____

LESSON 9-3 Reading Strategies
Vocabulary Development

The **mean**, the **median**, and the **mode** are measures that tell about the middle part of a set of data.
This chart will help you learn about each of them.

Mean (average)
The sum of all values divided by total number of values

Median
The middle number, or the average of the two middle numbers

Measures of the Middle Part of a Set of Data

Mode
The value or values that occur most often

Use the chart to answer each question.

1. What is the median?
   _____
   _____

2. What is the mean?
   _____

3. What is a mode?
   _____

The following list shows the number of goals scored each month by a hockey team.
4   4   7   6   8   3   5

Answer each question.

4. What is the mode of the data?

5. What is the median of the data?

Copyright © by Holt, Rinehart and Winston.
All rights reserved.

Holt Mathematics

Name _____ Date _____ Class _____

LESSON 9-3 Puzzles, Twisters & Teasers
Math a la Mode!

Find the mean, median, or mode for each data set. Each answer has a corresponding letter. Use the letters to solve the riddle.

Find the mean.

1. 20, 17, 42, 26, 27, 12, 31 _____ E
2. 15, 10, 12, 10, 13, 13, 10, 3 _____ S
3. 8, 5, 3, 75, 7, 3, 4, 7, 9, 2, 8, 5, 7 _____ T
4. 4.0, 3.3, 5.6, 4.6, 3.3, 5.6 _____ P

Find the median.

5. 35, 21, 34, 44, 36, 42, 29 _____ A
6. 23, 13, 45, 56, 72, 44, 89, 92, 67 _____ L
7. 5, 2, 12, 7, 13, 9, 8 _____ O
8. 4.3, 1.3, 4.5, 8.6, 9, 3, 2.1 _____ H

Find the mode.

9. 92, 88, 84, 86, 88 _____ C
10. 6, 8, 6, 7, 9, 2, 4, 22 _____ I
11. 7, 5, 4, 6, 8, 3, 5, 2, 5 _____ T
12. 2.0, 4.5, 6.2, 3.2, 4.5, 6.5 _____ N

Why did the elephant sit on the marshmallow?

It didn't want to

10.75	4.4	56	35	10.75	4.3
6	4.5	5	8		
11	4.3	25			
4.3	8	5			

| 88 | 4.3 | 8 | 88 | 8 | 56 | 35 | 11 | 25 |

Name _____ Date _____ Class _____

Practice A
LESSON 9-4 Variability

Find the least value, greatest value, and median for each data set.

1. 6, 9, 3, 7, 8, 7, 5

   least value: _____

   greatest value: _____

   median: _____

2. 12, 8, 24, 19, 15, 20, 13

   least value: _____

   greatest value: _____

   median: _____

Find the given values for each data set. Then use the values to make a box-and-whisker plot.

3. 27, 33, 28, 26, 34, 40, 21

   least value: _____

   greatest value: _____

   median: _____

   first quartile: _____

   third quartile: _____

   ←—+——+——+——+——+——+——+——+——+——+——→

4. 48, 64, 49, 55, 67, 50, 35, 62, 44, 52, 58

   least value: _____

   greatest value: _____

   median: _____

   first quartile: _____

   third quartile: _____

   ←—+——+——+——+——+——+——+——+——+——+——+——→

Copyright © by Holt, Rinehart and Winston.
All rights reserved.

Holt Mathematics

Name _____ Date _____ Class _____

LESSON 9-4 Practice B
Variability

Find the first and third quartiles for each data set.

1. 37, 48, 56, 35, 53, 41, 50

   first quartile: _____

   third quartile: _____

2. 18, 20, 34, 33, 16, 44, 42, 27

   first quartile: _____

   third quartile: _____

Use the given data to make a box-and-whisker plot.

3. 55, 46, 70, 36, 43, 45, 52, 61

4. 23, 34, 31, 16, 38, 42, 45, 30, 28, 25, 19, 32, 53

Use the box-and-whisker plots to compare the data sets.

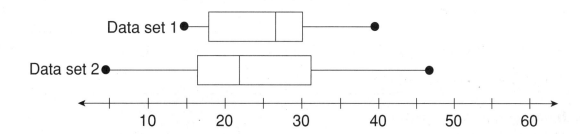

5. Compare the medians and ranges.

   _____

   _____

6. Compare the ranges of the middle half of the data for each set.

   _____

   _____

Name _____ Date _____ Class _____

LESSON 9-4 Practice C
Variability

Use the given data to make a box-and-whisker plot.

1. 76, 53, 55, 64, 43, 67, 73, 82, 71, 49, 58, 64

2. 2.1, 2.8, 3.4, 5.2, 4.3, 3.8, 3.0

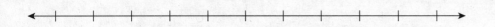

Use the box-and-whisker plots to compare the data sets.

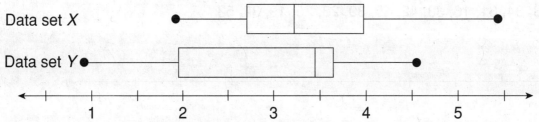

3. Compare the medians and ranges.

   _____

   _____

4. Compare the ranges of the middle half of the data for each set.

   _____

Match each set of data with a box-and-whisker plot.

5. range: 32; first quartile: 17; third quartile: 34 _____

6. range: 22; first quartile: 17; third quartile: 29 _____

7. range: 23; first quartile: 12; third quartile: 25 _____

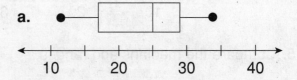

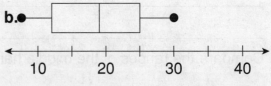

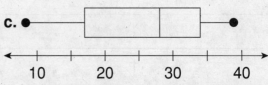

Name _____ Date _____ Class _____

LESSON 9-4 Reteach
Variability

Quartiles divide a data set into four equal parts.

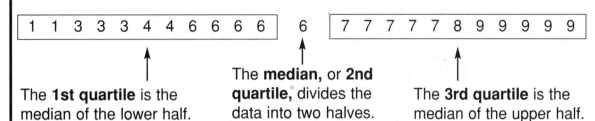

The **1st quartile** is the median of the lower half.

The **median**, or **2nd quartile**, divides the data into two halves.

The **3rd quartile** is the median of the upper half.

For each data set, circle and label the median *M*. Circle and label the first quartile Q_1. Circle and label the third quartile Q_3.

1. 3   3   3   4   5   6   7

2. 3   3   3   4   4   5   6   7   9   10   10

A **box-and-whisker** plot displays the quartile values as well as the lowest and highest numbers of a data set. The sides of the box are the first and third quartiles; the median is inside the box. The whiskers connect the box to the lowest and highest values.

This plot displays the values for the data set at the top of this page.

lowest value = 1
1st quartile = 4
median = 6
3rd quartile = 8
highest value = 9

Complete to make a box-and-whisker plot for the data set 45, 47, 47, 48, 48, 49, 53.

3. First, calculate three significant values for the data set.

median = _____   1st quartile = _____   3rd quartile = _____

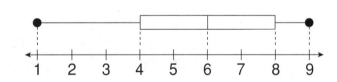

Holt Mathematics

Name _____ Date _____ Class _____

LESSON 9-4 Challenge
What's Normal?

Standard deviation (symbol σ, *sigma*) is a measure of variability that tells how far data are spread out from the mean of a data set.

In many situations, such as scores on the SAT or other standardized tests, the data cluster around the mean in such a way that if they are graphed to show the frequency of measures, the graph appears as a **bell-shaped curve**, also called the **normal curve**.

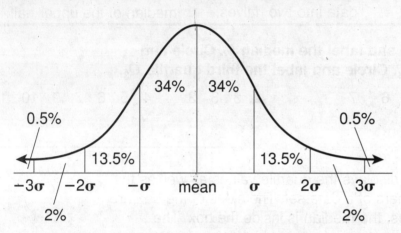

If the mean math score for males on the 2001 SAT I was 533 and the standard deviation was 115, determine the scores achieved by about 68% of the male participants.

According to the normal curve, 68% of scores fall between −σ and σ.
mean − σ = 533 − 115 = 418   mean + σ = 533 + 115 = 648
So, the scores for about 68% of the males fell between 418 and 648.

Assume a normal distribution for each situation.

1. A survey of 16-year-olds showed that they watched an average (mean) of 9.4 hours of TV per week, with a standard deviation of 1.2 hours. Determine how many hours of TV were watched by about:

   a. 68% of the participants. _____

   b. 95% of the participants. _____

2. On a certain standardized test, the mean score was 50 and the standard deviation 3. About what percent of the participants scored:

   a. between 50 and 56? _____

   b. 44 and 47? _____

Name _____ Date _____ Class _____

LESSON 9-4 Problem Solving
Variability

Write the correct answer.

1. Find the median of the data.

   _____

2. Find the first and third quartiles of the data.

   _____

3. Make a box-and-whisker plot of the data.

Super Bowl Point Differences

Year	Point Difference
2001	27
2000	7
1999	15
1998	7
1997	14
1996	10
1995	23
1994	17
1993	35
1992	13

The box-and-whisker plots compare the highest recorded Fahrenheit temperatures on the seven continents with the lowest recorded temperatures. Choose the letter for the best answer.

4. Which statement is true?

   A The median of the high temperatures is less than the median of the low temperatures.

   B The range of low temperatures is greater than the range of high temperatures.

   C The range of the middle half of the data is greater for the high temperatures.

   D The median of the high temperatures is 49°F.

5. What is the median of the high temperatures?

   F 128°F **H** −67°F
   G 120°F **J** −90°F

6. What is the range of the low temperatures?

   A 77°F **C** 120°F
   B 79°F **D** 129°F

Name _____ Date _____ Class _____

LESSON 9-4 Reading Strategies
Analyze Information

One way to organize data is to divide it into four equal parts called **quartiles**.

Here are the ages of students in the school chorus:
12 12 12 13 13 13 14 14 15 15 15

To organize this data into **quartiles**:

• First, find the **median**, the value that divides the set of data in half.

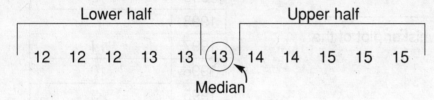

1. What value divides the data in half?

   _____

• The first and second quartiles are divided by the middle value of data *below* the median.

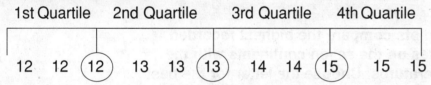

2. What is the middle value below the median?

   _____

• The third and fourth quartiles are divided by the middle value of the data *above* the median.

3. What is the middle value above the median?

   _____

4. When a group of data is organized into four equal parts, what is each part called?

   _____

Circle the quartile in which each age is located.

5. 12   1st quartile   2nd quartile   3rd quartile   4th quartile

6. 14   1st quartile   2nd quartile   3rd quartile   4th quartile

Name _____ Date _____ Class _____

Puzzles, Twisters & Teasers
LESSON 9-4 *What Are Your Values?*

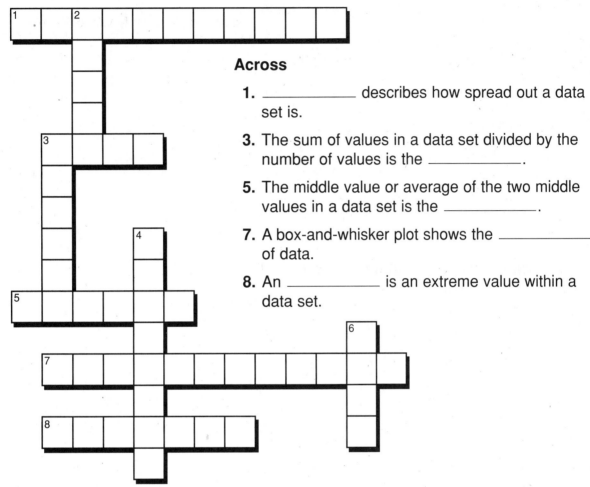

Across

1. _____ describes how spread out a data set is.
3. The sum of values in a data set divided by the number of values is the _____.
5. The middle value or average of the two middle values in a data set is the _____.
7. A box-and-whisker plot shows the _____ of data.
8. An _____ is an extreme value within a data set.

Down

2. The _____ of a data set is the largest value minus the smallest value.
3. A measure of central tendency describes the _____ of a data set.
4. A _____ divides a data set into four equal parts.
6. The value or values that occur most often in a data set is the _____.

Holt Mathematics

Name _____ Date _____ Class _____

LESSON 9-5 Practice A
Displaying Data

1. Use the data to complete the double-bar graph.

Softball Scores	1	2	3	4	5	6	7	8
Team A	3	2	0	2	1	4	2	1
Team B	1	4	3	0	3	1	2	1

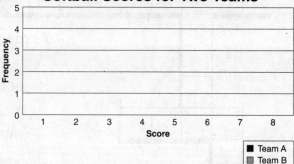

2. Use the data to make a histogram with intervals of 10.

Average High Temperatures in April in Tourist Cities				
Acapulco, Mexico	87	Montreal, Canada	51
Athens, Greece	67	Nassau, Bahamas	81
Dublin, Ireland	54	Paris, France	60
Hong Kong, China	79	Rome, Italy	68
London, U.K.	56	Sydney, Australia	73
Madrid, Spain	63	Toronto, Canada	51

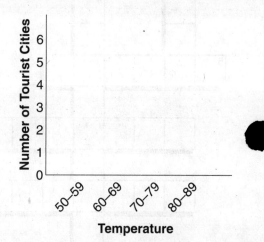

3. Make a double-line graph of the given data. Use the graph to estimate the heights of Dean and Susan when they were 9 years old.

Age	Dean's Height	Susan's Height
2	35 in.	30 in.
4	41 in.	37 in.
6	46 in.	44 in.
8	50 in.	51 in.
10	57 in.	58 in.
12	60 in.	65 in.

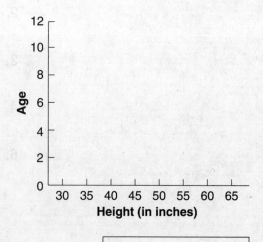

At 9 years old, Dean was approximately _____ tall, and Susan was approximately _____ tall.

Name _____ Date _____ Class _____

LESSON 9-5 Practice B
Displaying Data

1. Make a double-bar graph.

Daily Hours Worked	6	7	8	9	10	11	12
Crew A	4	3	6	1	3	1	2
Crew B	5	5	4	3	2	0	1

Daily Hours Worked by Two Crews

Frequency

Hours Worked

2. Use the data to make a histogram with intervals of 5.

Weekly Allowance of 20 Students			
$5	$15	$2	$10
$12	$12	$10	$15
$10	$5	$6	$4
$8	$7	$20	$7
$5	$4	$5	$9

Number of Students

Allowance (dollars)

3. Make a double-line graph of the given data. Use the graph to estimate the number of radio stations and cable TV systems in 2002.

Commercial Media in the United States		
Year	Radio Stations	Cable TV Systems
1997	10,207	10,950
1999	10,444	10,700
2001	10,516	9,924
2003	10,605	9,339

U.S. Commercial Media

Number of Enterprises

Year

Copyright © by Holt, Rinehart and Winston.
All rights reserved.

Holt Mathematics

Name _____ Date _____ Class _____

LESSON 9-5 Practice C
Displaying Data

1. Maggie found hotels in New York City offering the following rates per night. Make a histogram of the rates with intervals of $50.

   $175 $134 $119 $275 $318

   $155 $99 $178 $202 $199

   $280 $160 $108 $137 $151

   $221 $299 $97 $148 $283

2. Make a double-line graph of the given data. Use the graph to answer questions 3 and 4.

Railroad Ridership (in millions)										
	1991	1992	1993	1994	1995	1996	1997	1998	1999	2000
Amtrack	22.0	21.3	22.1	21.2	20.7	19.7	20.2	21.1	21.5	22.5
Commuter	18.1	20.3	32.9	39.5	42.2	45.9	48.5	54.0	58.3	61.6

Railroad Ridership (millions)

Year

3. Do you predict the number of riders increased or decreased in 2001?

4. Between which years did the number of commuter train riders increase the most?

Name _____ Date _____ Class _____

LESSON 9-5 Reteach
Displaying Data

A double-bar graph is used to compare quantities measured in the same unit.

The bars may be vertical.   The bars may be horizontal.

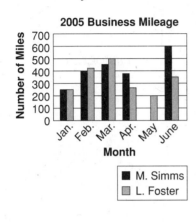
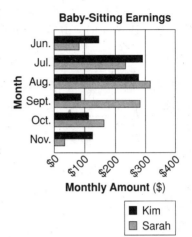

For Exercises 1-3, refer to the table below.

League Champion Batting Averages

Year	1997	1998	1999	2000	2001
National League	0.372	0.363	0.379	0.372	0.350
American League	0.347	0.339	0.357	0.372	0.350

1. If the horizontal axis will show the years and the vertical axis will show the batting averages, will you use vertical bars or horizontal bars for the graph?

2. To avoid decimals for the vertical scale, include *(in thousandths)* as part of the label. Then, what are appropriate values for the vertical scale?

3. Draw a double-bar graph for the data.

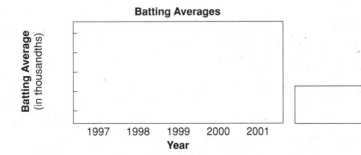

39   Holt Mathematics

Name _____ Date _____ Class _____

LESSON 9-5 Reteach
Displaying Data (continued)

A **histogram** represents intervals of grouped data as bars with no space between. The data may be organized in a **frequency table**.

The lengths of selections most requested today by listeners of station WXYZ are shown in a frequency table and a histogram.

Frequency Table

Selection Length (in seconds)	Frequency (number of requests)
60–119	2
120–179	5
180–239	12
240–299	14
300–359	7
360–419	2

Histogram

Complete to make a frequency table and histogram for this data set.

The heights, in inches, of 25 students in a high school class are 70, 72, 59, 56, 55, 60, 48, 72, 59, 48, 66, 72, 58, 60, 60, 50, 68, 72, 68, 62, 72, 58, 60, 68, 59.

4. Complete this frequency table for the data.

Interval	Tallies	Frequency
70–74	⊞ I	6
65–69		
60–64		
55–59		
50–54		
45–49		

5. Use the frequency table to make a histogram.

6. Discuss how the histogram for this data would change if the intervals were 40–49, 50–59, 60–69, 70–79.

_____

_____

_____

Name _____ Date _____ Class _____

Challenge
LESSON 9-5 Double the Fun

A **double-bar graph** is used to compare between data groups and within data groups.

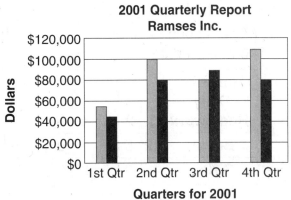

2001 Quarterly Report Ramses Inc.

☐ Income
■ Expenses

1. In which quarter did the company experience its greatest profit? Explain.

   _____

   _____

2. Did the company experience a profit or loss for the year? Estimate the amount.

   _____

3. Make a double-bar graph to display this data.

Baseball Home-Run Leaders, years 1997–2001

	1997	1998	1999	2000	2001
American League	56	56	48	47	52
National League	49	70	65	50	73

4. In which year was the difference between home-run leaders greatest?

   _____

5. Compare the home-run leader data for the American League and the National League.

   _____

   _____

Name _____ Date _____ Class _____

LESSON 9-5 Problem Solving
Displaying Data

Make the indicated graph.

1. Make a double-bar graph of the homework data.

Hours of Daily Homework	1	2	3	4	5
Boys	12	5	2	1	0
Girls	4	6	5	3	2

Hours of Daily Homework

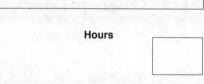

Hours

2. The annual hourly delay per driver in the 20 U.S. cities with the most traffic are as follows: 56, 42, 53, 46, 34, 37, 42, 34, 53, 21, 45, 50, 34, 42, 41, 38, 42, 34, 38, 31. Make a histogram with intervals of 5 hours.

For 3–5, refer to the double-line graph. Circle the letter of the correct answer.

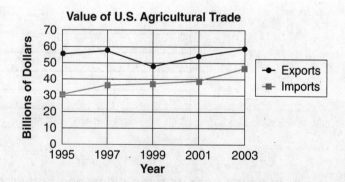

3. Estimate the value of U.S. agricultural exports in 1998.
   A $62 billion
   B $59 billion
   C $52 billion
   D Cannot be determined

4. Estimate the value of U.S. agricultural imports in 2000.
   F $39 billion
   G $31 billion
   H $29 billion
   J $21 billion

5. Estimate the difference between agricultural exports and imports in 1995.
   A $16 billion
   B $21 billion
   C $26 billion
   D Cannot be determined

Name _____ Date _____ Class _____

Reading Strategies
LESSON 9-5 Use Graphic Aids

A **line graph** shows changes in data over time.
- The temperature changes from morning to night.
- Your age changes from year to year.

This double-line graph shows money earned from ice cream and hot dogs sold during a class picnic.

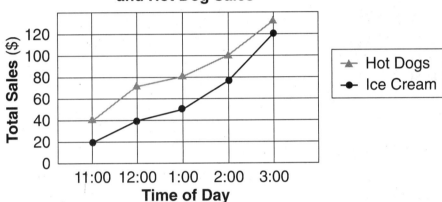

Use the line graph to answer each question.

1. What information is found along the left side of the graph?

   _____

2. The values on the left side of the graph increase by what amount?

   _____

3. What information is shown along the bottom of the graph?

   _____

Each point on the graph shows total sales up to that time.

4. How much money was made from ice cream and hot dog sales by 11:00 A.M.?

   _____

5. During which time period was the least amount of ice cream sold?

   _____

Puzzles, Twisters & Teasers
LESSON 9-5 — Cold Facts!

Find and circle the words from the list in the word search (horizontally, vertically or diagonally). Find a word that answers the riddle. Circle it and write it on the line.

bar   graph   frequency   table   histogram
line display   organize   data   random

```
Q T Y H B G U J N M I K O P
W I N D O W S D A T A V H A
R F T R R U I I N A N J I S
A P S D G F T S E B A R S D
N T G B A R D P I U Y T T F
D Z X C N W S L I N E A O B
O N M K I G R A P H J B G N
M L O I Z Q W Y V F T L R A
V G F R E Q U E N C Y E A Y
Q A Z X S W E D C V F R M X
```

Why did the computer catch a cold?

It forgot to close its _____.

Name _____ Date _____ Class _____

LESSON 9-6 Practice A
Misleading Graphs and Statistics

Explain why each graph is misleading.

1. The Price of a Pound of Apples in Selected Cities in September 2000

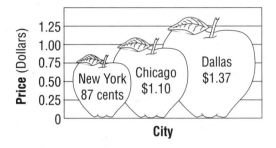

2. Radio Formats People Listen to Most

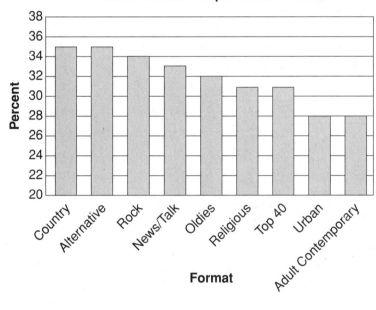

_____
_____
_____

Explain why the statistic is misleading.

3. A juice company surveyed 4 people about which juice they preferred. Three of the people preferred the company's juice over the competition's. The company published that 3 times more people preferred their juice.

_____

45 Holt Mathematics

Name _____ Date _____ Class _____

LESSON 9-6 Practice B
Misleading Graphs and Statistics

Explain why each graph is misleading.

1. **On the Road**
Number of Trucks that Travel City Roads

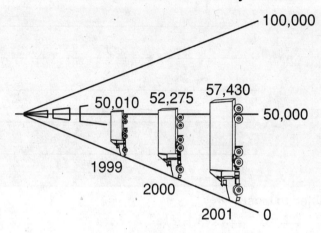

_____
_____
_____
_____
_____
_____
_____

2. **Federal Minimun Wage Rates Since 1980**

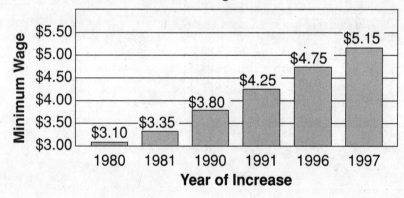

_____
_____
_____

Explain why the statistic is misleading.

3. A chewing gum company advertises that the flavor of its new chewing gum lasts for an average of 55 minutes based on the following durations reported by customers: 12 min, 33 min, 5 min, 200 min, and 25 min.

_____
_____

Name _____ Date _____ Class _____

Practice C
LESSON 9-6 Misleading Graphs and Statistics

Explain why each graph is misleading.

1. **Number of Units Sold**

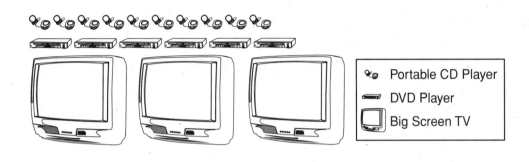

_____
_____

2.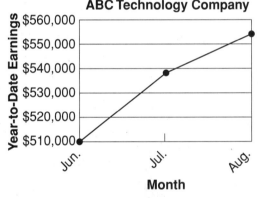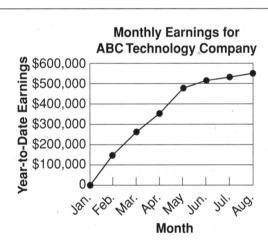

_____
_____
_____

Explain why the statistic is misleading.

3. A hotel advertises that its average monthly temperature is 78°F. The average temperature for each month is 52°F, 57°F, 59°F, 90°F, 92°F, 98°F, 104°F, 95°F, 94°F, 93°F, 53°F, and 49°F.

_____
_____

Name _____ Date _____ Class _____

LESSON 9-6 Reteach
Misleading Graphs and Statistics

The most appropriate measure of central tendency should be used.

The scale of graphs should begin at 0, or use a broken scale.

Complete to tell why the situation is misleading.

When Harold got 70 on his math test, he told his family that the mode for the test was 70. He said, "More students got 70 than any other grade." The actual grades, which he did not tell his family, were 70, 70, 70, 81, 85, 86, 90, 94, 95, 96, 97, 100.

1. How did Harold mislead his family?

   _____

2. What was the mean of the class? the median?

   _____

3. What would have been a more accurate way for Harold to tell his family how his grade compared with the other grades?

   _____

Complete to tell why the situation is misleading.

The graph shows the salaries at a small company.

4. Use the heights of the bars to estimate the ratio of the salaries of Ms. C to Ms. D.

   _____

5. According to the scale, what is Ms. C's salary? Ms. D's salary?

   _____

6. What is the actual ratio of their salaries?

   _____

7. Why is the graph misleading?

   _____

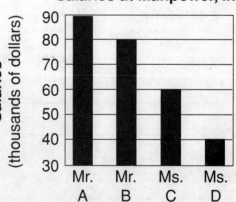
Salaries at Manpower, Inc.

Name _____ Date _____ Class _____

LESSON 9-6 Challenge
To Tell the Truth...

For each situation, a statement based on the data is made. The statement is misleading. Explain why.

1. Ten students took the SAT I in their junior year. Then all ten took a course to prepare for the test the following year. Their average score on the math section increased from 525 to 565. The company offering the course stated in their new ads, "Students who took our course added 40 points to their math SAT."

   _____

   _____

2. "Don't worry, Jack. You can dive here. The average depth of the pool is 8 ft."

   _____

   _____

   _____

3. Bill's test scores: 50, 52, 55, 63, 71, 98 Maria's test scores: 50, 85, 87, 92, 93, 98 "Both students' scores range from 50 to 98."

   _____

   _____

4. An ad says "Save 50% on all regular price clothing! Take 20% off of all regular price clothing! Then take 30% of the newly reduced price for a total savings of 50%!"

   _____

   _____

5. A reporter asked twelve people if they went to the movies last week. Of the five who answered yes, he asked how many times they had gone to the movies. The responses were: 1, 1, 2, 2, and 4. The reporter then wrote "The average person attended two movies last week."

   _____

Problem Solving
LESSON 9-6: Misleading Graphs and Statistics

Explain why each statistics is misleading.

1. A poll taken at a college says that 38% of students like pizza the best, 32% like hamburgers the best, and 30% like spaghetti the best. They conclude that most of the students at the college like pizza the best.

   _____
   _____

2. The National Safety Council of Ireland found that young men were responsible in 57% of automobile accidents they were involved in. The NSC Web site made this claim: "Young men are responsible for over half of all road accidents."

   _____
   _____

3. Explain why the Centers for Disease Control (CDC) has been highly criticized for the graph below.

   _____
   _____
   _____

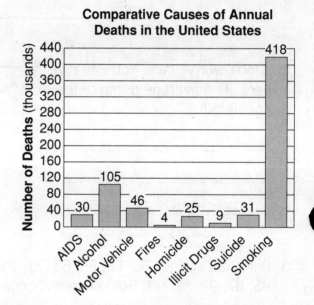

Comparative Causes of Annual Deaths in the United States

Choose the letter for the best answer.

4. Which statement is a misleading statistic for the data in the table?

Student	Test Grade
A	85%
B	92%
C	88%
D	10%
E	80%

   A The median score was 85%.
   B Most students scored an 80% or above.
   C The average test score was 71%.
   D The range of test scores was 82.

5. A sno-cone store claims, "Our sales have tripled!" Sno-cone sales from March to May were 50 and sales from June to August were 150. Why is this misleading?

   A Sample size is too small.
   B During the summer, sales should be higher.
   C Should use the median not mean.
   D The statement isn't misleading.

Name _____ Date _____ Class _____

Reading Strategies
9-6 Compare and Contrast

Graph A and Graph B are bar graphs of the same data. A graph can be misleading if the numbers on the left side of the graph do not start at 0.

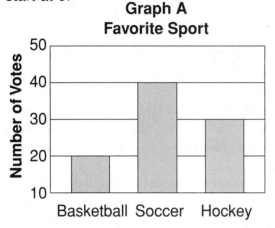
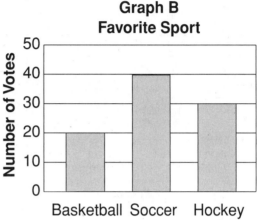

Answer each question.

1. Compare the titles of Graph A and Graph B.

   _____

2. By how many does the number of votes increase on both graphs?

   _____

3. Compare the first number shown along the left side of Graph A and the first number shown along the left side of Graph B.

   _____

   _____

4. Compare the bars on Graph A to Graph B.

   _____

5. Which graph is misleading? Why?

   _____

   _____

Name _____ Date _____ Class _____

LESSON 9-6 Puzzles, Twisters & Teasers
Truth or Consequences!

Use the two graphs to answer the questions. Match the letters to your answers to solve the riddle.

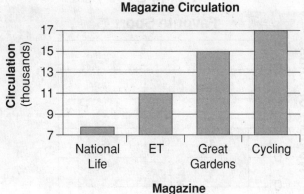

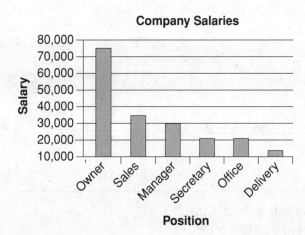

1. What is the scale on the magazine circulation graph? _____ **S**

2. What is the scale on the company salaries graph? _____ **P**

3. The circulation of Great Gardens is how many times greater than that of National Life? _____ **M**

4. The owner's salary appears to be how many times greater than the secretary's? _____ **U**

5. The circulation of Cycling is how many thousands more than ET? _____ **P**

What shoes should you wear when your basement is flooded?

___ ___ ___ ___ ___
10,000   5   2   6   2000

Name _____ Date _____ Class _____

Practice A
9-7 Scatter Plots

1. Use the given data to make a scatter plot.

Calories and Fat Per Portion of Meat & Fish

	Fat (grams)	Calories
Fish sticks (breaded)	3	50
Shrimp (fried)	9	190
Tuna (canned in oil)	7	170
Ground beef (broiled)	10	185
Roast beef (relatively lean)	7	165
Ham (light cure, lean and fat)	19	245

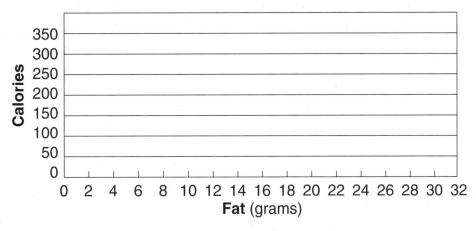

Calories and Fat Per Portion of Meat and Fish

Do the data sets have a positive, a negative, or no correlation?

2. The size of the bag of popcorn and the price of the popcorn

3. The increase in temperature and number of snowboards sold

4. Use the data to predict how much money Tyler would be paid for babysitting $7\frac{1}{2}$ hours.

Amount Tyler Earns Babysitting

Hours	1	2	3	4	5	6	7	8
Amount	$4	$8	$12	$16	$20	$24	$28	$32

According to the data, Tyler would get paid $_____ for babysitting $7\frac{1}{2}$ hours.

Practice B
9-7 Scatter Plots

1. Use the given data to make a scatter plot.

Tall Buildings in U.S. Cities

Building	City	Stories	Height (meters)
Sears Tower	Chicago	110	442
Empire State Building	New York	102	381
Bank of America Plaza	Atlanta	55	312
Library Tower	Los Angeles	75	310
Key Tower	Cleveland	57	290
Columbia Seafirst Center	Seattle	76	287
NationsBank Plaza	Dallas	72	281
NationsBank Corporate Center	Charlotte	60	265

Tall Buildings in U.S. Cities

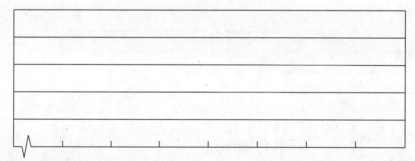

Do the data sets have a positive, a negative, or no correlation?

2. The temperature outside and the number of ice cream cones sold

3. The amount of time spent in the bathtub and the temperature of the bath water

4. Use the data to predict the percent of Americans owning a home in 1955.

Percent of Americans Owning Homes

Year	1950	1960	1970	1980	1990
Percent	55.0%	61.9%	62.9%	64.4%	64.2%

According to the data, about _____ % of Americans owned a home in 1955.

Name _____ Date _____ Class _____

LESSON 9-7 Practice C
Scatter Plots

1. Use the given data to make a scatter plot of length and maximum depth of the largest lakes of North America.

Largest Lakes of North America

Lake	Area (square miles)	Length (miles)	Maximum Depth (ft)
Superior	31,700	350	1330
Huron	23,000	206	750
Michigan	22,300	307	923
Great Slave	11,031	298	2015
Erie	9910	241	210
Ontario	7340	193	802
Athabasca	3064	208	407
Reindeer	2568	143	720

Largest Lakes of North America

[blank grid for scatter plot]

Do the data sets have a positive, a negative, or no correlation?

2. The amount of time spent exercising and the number of calories burned

   _____

3. The discount percent of an item and the total cost of the item

   _____

4. Use the data to estimate the median four-person family income in 1982.

Median Four-Person Family Income

Year	1975	1980	1985	1990	1995	1999
Income	$15,848	$24,332	$32,777	$41,151	$49,687	$59,981

According to the data, the median four-person family income was about _____ in 1982.

Name _____ Date _____ Class _____

LESSON 9-7 Reteach
Scatter Plots

Two sets of data can be graphed as points in a **scatter plot**.
If there is a relationship between the data sets, a **line of best fit**
can be drawn.

Positive correlation:
both sets of data
increase together.

Negative correlation:
values of one set
increase while values of
the other set decrease.

No correlation: points
neither increase nor
decrease together.

  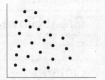

Make a scatter plot. Include a line of best fit if there is a
correlation. Describe the correlation.

1.

Time (hours)	1	2	2.5	6
Distance (miles)	50	150	175	270

The values for
time are: increasing

distance are: _____

So, there is a ____ _____ correlation.

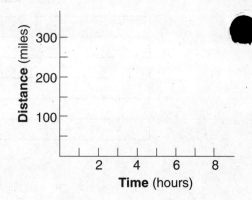

2.

Number of Workers	6	4	2	1
Number of Days	1	2	5	7

The values for
    the number of workers are: _____
    the number of days are: _____
So, there is a _____ _____ correlation.

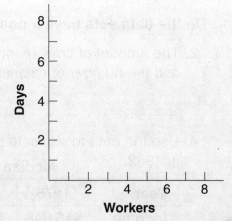

Copyright © by Holt, Rinehart and Winston.
All rights reserved.

Holt Mathematics

Name _____ Date _____ Class _____

Challenge
LESSON 9-7
This Fits Nicely!

When two sets of data show a correlation, you can draw a **line of best fit** that approximates a trend.

Here are some data relating the gestation periods of selected animals to their average life spans. The data are separated into 3 equal sets.

	Set I			Set II			Set III		
Gestation (days)	31	61	68	105	151	167	285	330	365
Longevity (years)	13	12	4	5	8	20	15	20	12

1. Determine the **median-median point** for each set of points by getting the median value for the gestation values and the median value for the longevity values.

   The median-median point for
   Set I is: _____   Set II is: _____   Set III is: _____

2. Make a scatter plot for the given data. Describe the correlation. _____

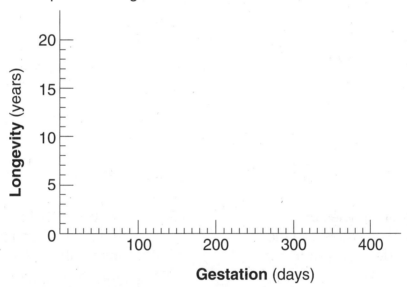

3. Using an X for each, plot the three median-median points on your graph.

   Using a ruler, draw a dotted line through the median-median points for Sets I and III.

   Keeping the ruler at the level of the dotted line, estimate the vertical distance between the dotted line and the median-median point for Set II. Then slide the ruler down about one-third this distance. Draw a solid line parallel to the dotted line. This solid line is called the median-median line and it is a line of best fit for the given data.

Problem Solving
9-7 Scatter Plots

Use the data given at the right.

1. Make a scatter plot of the data.

Percent of Americans Who Have Completed High School

Year	Percent
1910	13.5
1920	16.4
1930	19.1
1940	24.5
1950	34.3
1960	41.1
1970	55.2
1980	68.6
1990	77.6
1999	83.4

2. Does the data show a positive, negative or no correlation?

3. Use the scatter plot to predict the percent of Americans who will complete high school in 2010.

Choose the letter for the best answer.

4. Which data sets have a positive correlation?

   A The length of the lines at amusement park rides and the number of rides you can ride in a day

   B The temperature on a summer day and the number of visitors at a swimming pool

   C The square miles of a state and the population of the state in the 2000 census

   D The length of time spent studying and doing homework and the length of time spent doing other activities

5. Which data sets have a negative correlation?

   F The number of visitors at an amusement park and the length of the lines for the rides

   G The amount of speed over the speed limit when you get a speeding ticket and the amount of the fine for speeding

   H The temperature and the number of people wearing coats

   J The distance you live from school and the amount of time it takes to get to school

Name _____ Date _____ Class _____

LESSON 9-7 Reading Strategies
Vocabulary Development

A **scatter plot** shows whether two sets of data are related. The data are shown as points.
This scatter plot shows data about a child's age and his or her shoe size. By looking at both, you can see if there is a relationship between the two sets of data.

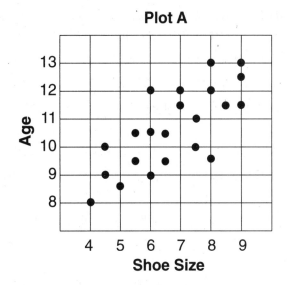

Answer each question.

1. What information is shown along the left side of the plot?

   _____

2. What is the age range of children shown on this plot?

   _____

3. What information is listed along the bottom of the plot?

   _____

4. What is the range of shoe sizes shown on this plot?

   _____

5. Describe the pattern of points on the plot.

   _____

6. As a child's age increases, what happens to his or her shoe size?

   _____

7. Is there a relationship between a child's age and his or her shoe size?

   _____

Holt Mathematics

Puzzles, Twisters & Teasers
9-7 The Plot Thickens!

Identify each pair of data sets as having a positive correlation, negative correlation or no correlation. Use the letters of your answers to solve the riddle.

1. The number of pages you have read in a book and the number of pages remaining
   T Positive Correlation   **O** Negative Correlation   **D** No Correlation

2. The day of the month and the wind speed
   T Positive Correlation   **O** Negative Correlation   **D** No Correlation

3. The age of a car and its selling price
   O Positive Correlation   **T** Negative Correlation   **D** No Correlation

4. The weight of a vehicle and its gas mileage
   L Positive Correlation   **M** Negative Correlation   **P** No Correlation

5. The outside temperature and the number of people in attendance at the beach
   Y Positive Correlation   **L** Negative Correlation   **S** No Correlation

6. The month of the year and the number of birthdays in a certain month
   S Positive Correlation   **O** Negative Correlation   **W** No Correlation

7. The population of a state and the number of senators
   L Positive Correlation   **R** Negative Correlation   **A** No Correlation

8. The length of your hair and the number of days since your last haircut
   N Positive Correlation   **T** Negative Correlation   **K** No Correlation

9. The number of hours spent studying and the test score received
   R Positive Correlation   **S** Negative Correlation   **B** No Correlation

Which two days of the week start with T?

T O D A Y A N D
3 1 2 7 5 7 8 2

T O M O R R O W
3 1 4 1 9 9 1 6

Name _____ Date _____ Class _____

LESSON 9-8 Practice A
Choosing the Best Representation of Data

1. Which graph is a better display of the number of vacation days certain countries have in a year?

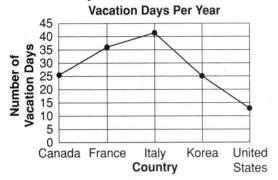
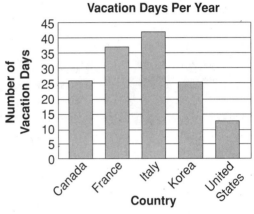

_____
_____

2. Which graph is a better display of the distribution of cats' weights?

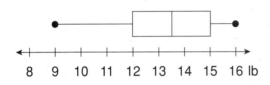
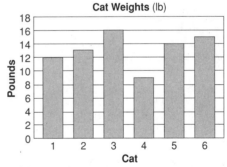

_____
_____

3. A scientist measured the lengths of 10 earthworms. The table shows her data. Choose an appropriate data display and draw the graph.

Earthworm Lengths (cm)				
8	14	9	10	9
11	7	10	12	9

Name _____ Date _____ Class _____

LESSON 9-8 Practice B
Choosing the Best Representation of Data

1. Which graph is a better display of the number of students in a class who chose math as their favorite subject?

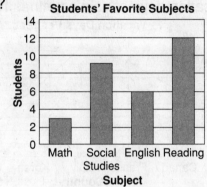

_____
_____

2. Which graph is a better display of the change in the number of cell telephone subscribers?

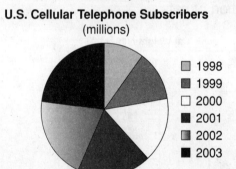
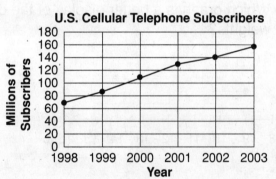

_____
_____

3. The table shows the heights of players on a school basketball team. Choose an appropriate data display and draw the graph.

Heights of Basketball Players (in.)			
70	64	68	71
61	68	65	73

Copyright © by Holt, Rinehart and Winston.
All rights reserved.

62

Holt Mathematics

Name _____ Date _____ Class _____

LESSON 9-8 Practice C
Choosing the Best Representation of Data

1. Which graph is a better display of the frequency of students' scores on a math quiz?

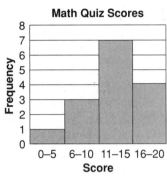
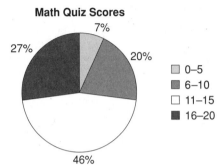

_____

_____

2. Which graph is a better display of the U.S. population after the last five censuses?

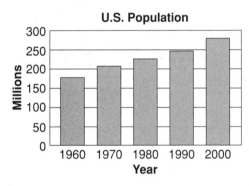
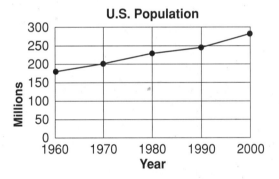

_____

_____

3. The table shows the ages and heights of corn plants. Choose an appropriate data display and draw the graph.

Age (days)	4	6	3	8	6	9	4	5	7	9
Height (cm)	3	4	3	7	5	8	4	7	4	6

63 Holt Mathematics

Name _____ Date _____ Class _____

LESSON 9-8 Reteach
Choosing the Best Representation of Data

A line graph is a good way to show changes over time. The amount of money in Tasha's account changes from month to month.

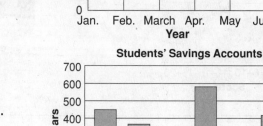
Tasha's Savings Account

A bar graph is a good way to compare different quantities. The amounts of money in the students' bank accounts are being compared.

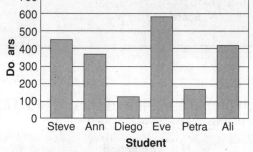
Students' Savings Accounts

A circle graph is a good way to compare parts of a whole. Lionel's assets are the whole; the parts are his savings account, checking account, and mutual funds.

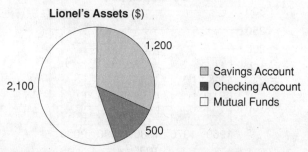
Lionel's Assets ($)

Determine which kind of graph would be best to display the described data. Write *line graph*, *bar graph*, or *circle graph*.

1. Total number of hits by each member of a softball team

   _____

2. Number of miles that a boy rides his bike each week

   _____

3. Daily profits of a hot dog stand at a county fair

   _____

4. Part of a hot dog stand's profits that come from popcorn sales

   _____

Name _____ Date _____ Class _____

LESSON 9-8 Challenge

What's Going On?

Invent a situation for the graph. Label the parts of the graph and describe the data it displays.

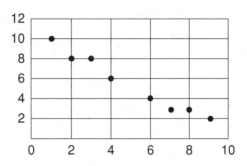

_____

_____

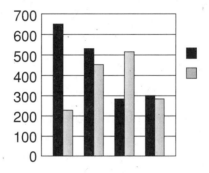

_____

_____

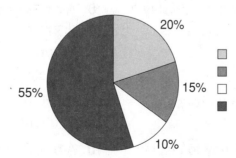

_____

_____

Name _____ Date _____ Class _____

Problem Solving
LESSON 9-8: Choosing the Best Representation of Data

Write what kind of graph would be best to display the described data.

1. Numbers of times that members of track team ran a mile in the following intervals: 4 min 31 s to 4 min 40 s, 4 min 41 s to 4 min 50 s, 4 min 51 s to 5 min, 5 min 1 s to 5 min 10 s
   _____

2. Distribution and range of students' scores on a history exam
   _____

3. Relationship between the amounts of time a student spent on her math homework and the numbers of homework problems she solved
   _____

4. Total numbers of victories of eight teams in an intramural volleyball league
   _____

5. Part of calories in a meal that come from protein
   _____

6. Numbers of books that a student reads each month over a year
   _____

Choose the letter for the best answer.

7. A bar graph is a good way to display
   A data that changes over time.
   B parts of a whole.
   C distribution of data.
   D comparison of different groups of data.

8. A circle graph is a good way to display
   F range and distribution of data.
   G distribution of data.
   H parts of a whole.
   J changes in data over time.

9. A scatter plot is a good way to display
   A comparison of different groups of data.
   B distribution and range of data.
   C the relationship between two sets of data.
   D parts of a whole.

10. A box-and-whisker plot is a good way to display
    F range and distribution of data.
    G the relationship between two sets of data.
    H data that changes over time.
    J parts of a whole.

Name _____ Date _____ Class _____

Reading Strategies
LESSON 9-8 *Understand Vocabulary*

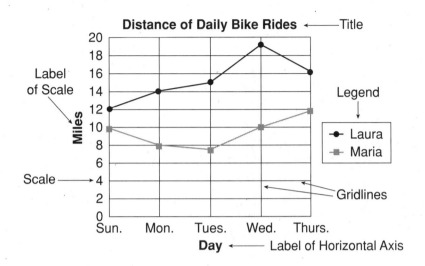

Refer to the graph at right.

1. What does the title of the graph tell about the graph?

   _____

   _____

2. What does the legend tell about the graph?

   _____

   _____

3. What does the scale of the graph show?

   _____

   _____

4. What is the range of the scale?

   _____

5. Which month did the team have the most losses?

   _____

6. Which month did the team have the most wins?

   _____

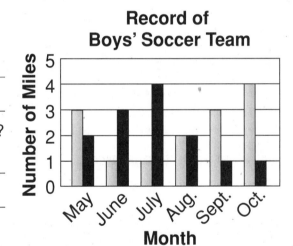

Name _____ Date _____ Class _____

LESSON 9-8 Puzzles, Twisters & Teasers
The Best Displays

Across

1. The kind of display preferred by cats
2. When you want to see the distribution
3. How a weightlifter might keep track of his progress

Down

1. Is there a relationship between the data sets?
2. Time changes everything
3. When you want to see who ate the biggest part of the pizza
4. What is the frequency?

Holt Mathematics

LESSON 9-1 Practice A
Samples and Surveys

Identify the sampling method used.

1. The name of each audience member at a game show is printed on a separate piece of paper and put into a rotating case. One audience member is chosen to play the game by drawing one piece of paper from the shuffled names.

   random

2. At a frozen pizza manufacturing plant, a coupon for a free pizza is put inside the package of every 100th pizza.

   systematic

3. The teacher asks that students with birthdays from July to December go to the chalkboard to work the next problem.

   systematic

Identify the population and sample. Give a reason the sample could be biased.

4. A radio disc jockey asks listeners to call in and name their favorite radio station.

   population _____ radio listeners
   sample _____ listeners who call in
   possible bias _____ listeners like that station

5. A hospital mails out surveys to 500 recent patients to get their feedback on their hospital visit.

   population _____ recent patients at the hospital
   sample _____ people who mail back survey
   possible bias _____ extremes more likely to respond

6. The first 10 people leaving the theater are asked to give their feedback about the movie.

   population _____ people who see the movie
   sample _____ moviegoers surveyed
   possible bias _____ early-exiters more likely to dislike the movie

LESSON 9-1 Practice B
Samples and Surveys

Identify the sampling method used.

1. People in the security line at the airport are asked to step out of the line for a more detailed search. The people pulled out of the line have not necessarily done anything wrong, and they are not chosen according to any particular rule.

   random

2. At the 1-mile marker of a marathon, a timekeeper shouts out the time elapsed to every 10th runner that passes by. A statistician records the times shouted.

   systematic

3. A geologist visits 10 randomly-selected lakes in the region and collects soil samples in randomly-selected areas along each shoreline.

   stratified

Identify the population and sample. Give a reason the sample could be biased.

4. At a convention of science teachers, various attendees are asked to name their favorite subject in high school.

   population _____ teachers at the convention
   sample _____ teachers surveyed
   possible bias _____ most will say science

5. Donors participating in a blood drive are given a small amount of money for their blood donation. Before they can give blood, each person is surveyed to find out if they are eligible to give blood.

   population _____ blood donors
   sample _____ blood donors (entire population)
   possible bias _____ people may lie to get money

6. Interviewers at the mall are surveying girls with red hair to find out if a correlation exists between personality and red hair.

   population _____ girls with red hair
   sample _____ girls surveyed
   possible bias _____ some girls color their hair

LESSON 9-1 Practice C
Samples and Surveys

Identify the sampling method used.

1. In clinical research done on a new medication, every other patient is given a placebo. (A placebo is an inactive substance that does not have the effect of the medicine. Placebos are used in controlled medical experiments.)

   systematic

2. A television station sends out reporters to randomly-selected neighborhoods throughout the city. In each neighborhood randomly-selected voters are polled to find the approval rating of the current mayor.

   stratified

3. A surveyor calls the last person on each page of a phone book.

   systematic

Identify the population and sample. Give a reason the sample could be biased.

4. The professional baseball league All-Star ballots are handed out at the stadium on game days.

   population _____ baseball fans
   sample _____ fans who attend games
   possible bias _____ some fans don't attend games

5. The city mayor visits a neighborhood council meeting and asks for a raise of hands to show support for his education agenda.

   population _____ people in neighborhood
   sample _____ people at meeting
   possible bias _____ none are against education

6. An online newspaper asks every person who reads a certain article to rate the article.

   population _____ people who visit the site
   sample _____ people who rate the article
   possible bias _____ readers interested in the topic more likely to rate it

LESSON 9-1 Reteach
Samples and Surveys

A *survey* uses a *small sample* to represent a *large population*.

Sampling Method	How Members of the Sample Are Chosen
Random	By chance; members have an equal chance of selection
Systematic	According to a rule or pattern
Stratified	At random from randomly-chosen subgroups

A senator's office sends workers to ask constituents at a local mall how they feel about floating a bond to acquire land around a reservoir to be preserved as open space. The population in this survey are all the eligible voters of the state. The shoppers at the mall are the sample of that population.

Identify the sampling method used.

1. A survey calls every 10th name listed in a local phone book.

   systematic

Identify the population and the sample.

2. Kennedy HS seniors planning to attend the prom were asked if senior dues should include a photo taken at the prom.

   Population _____ all seniors at Kennedy HS
   Sample _____ Kennedy seniors planning to attend the prom

A *biased sample* is not a good representative of the population.

A school principal asks parents attending an art workshop if funding for a theater arts program should be included in the school budget. This is a biased sample since the parents attending an art workshop are likely to be in favor of additional art programs.

Identify the population and the sample. Give a reason why the sample could be biased.

3. Homeowners within a 10-mile radius of a nuclear power plant were asked if they think the plant should be closed.

   Population _____ NY homeowners
   Sample _____ homeowners in a 10-mi radius of the plant
   Possible Bias: Homeowners close to the plant are more likely to want it closed.

Challenge
9-1 Stay in the Margins

A survey gathers information from a few people and then the results are used to reflect the opinions of a larger population. Researchers and pollsters use sample populations because it is cheaper and easier to poll a few people than to ask everybody. One key to successful surveys of sample populations is finding the appropriate size for the sample that will give accurate results without spending too much time or money.

Suppose that 900 American teens were surveyed about their favorite ski category of the 2002 Winter Olympics in Park City, Utah. Ski Jumping was the favorite for 20% of those surveyed. This result can be used to predict how many of all 31 million American teens favor Ski Jumping.

31,000,000 × 0.20 = 6,200,000 American teens favor Ski Jumping

To determine how accurately the results of surveying 900 American teens truly reflect the results of surveying all 31 million American teens, a **margin of error** should be given. When pollsters report the margin of error for their surveys, they are stating their confidence in the data they have collected.

The margin of error can be calculated by using the formula $\frac{1}{\sqrt{n}}$, where n is the number in a sample size.

For the above sample the margin of error would be $\frac{1}{\sqrt{900}} = \frac{1}{30} = 0.03\overline{3} \approx 3\%$. Since the actual statistic could be smaller or larger than the true amount, the margin of error is expressed as ±3%.

1. Find the margin of error for a survey of 100 American teens. ____±10%____
2. Compare that margin of error to the margin of error of 900 teens.

   __The survey of 900 teens is more accurate by ±7 percentage points.__

3. Find the margin of error for a survey of 9,000 American teens. ____±1.1%____
4. Find the margin of error for a survey of 90,000 American teens. ____±0.3%____
5. Draw a conclusion about the margin of error based on size of the sample.

   __Possible answer: The larger the sample, the smaller the margin of error.__
   __However, the decrease in the margin of error is not proportional with the__
   __increase in the sample size.__

Problem Solving
9-1 Samples and Surveys

Identify the sampling method used.

1. Every twentieth student on a list is chosen to participate in a poll.

   __Systematic__

2. Seat numbers are drawn from a hat to identify passengers on an airplane that will be surveyed.

   __Random__

Give a reason why the sample could be biased. Possible answers:

3. A company wants to find out how its customers rate their products. They ask people who visit the company's Web Site to rate their products.

   __Many people who use the prod-__
   __ucts will not visit the Web Site.__

4. A teacher polls all of the students who are in detention on Friday about their opinions on the amount of homework students should have each night.

   __Students in detention may be__
   __the least likely to do homework.__

A car dealership wants to know how people who have visited the dealership feel about the dealership and the sales people. They survey every 5th person who buys a car. Choose the letter for the best answer.

5. Identify the population.
   A) People who visit the dealership
   B) People who buy a car from the dealership
   C) People in the local area
   D) The salesmen at the dealership

6. Identify the sample.
   F) Every person who visits the dealership
   G) People who buy a car
   (H)) Every 5th buyer
   J) People in the local area

7. Identify the possible bias.
   A) Not all people will visit the dealership.
   B) Did not survey everyone who buys a car.
   (C)) Not including those who visited but did not buy.
   D) There is no bias.

8. Identify the sampling method used.
   F) Random
   (G)) Systematic
   H) Stratified
   J) None of these

Reading Strategies
9-1 Compare and Contrast

Surveys are taken to get information about a group of people. A survey may include:
- the entire group, called the **population**;
- part of the group, called a **sample** of the population.

Compare a population to a sample.

1. How are a sample and a population alike?

   __Possible answer: They are both groups of people being surveyed.__

2. How are a population and a sample different?

   __Possible answer: In a population survey, every member of the group is__
   __surveyed; a sample surveys only part of a group.__

Sometimes it is impossible to survey an entire population. Most of the time only a sample of the population is surveyed. There are two kinds of samples.
- An **unbiased sample** accurately represents the population.
- A **biased sample** does not represent the entire population.

A company wants to test market a new sports drink for high school students. Write "unbiased sample" or "biased sample" for each situation.

3. If the company surveys every tenth student in seven different high schools across the country, it would be a(n) __unbiased sample__

4. If the company surveys only athletes at the seven high schools, it would be a(n) __biased sample__

Compare an unbiased sample to a biased sample.

5. How is an unbiased sample like a biased sample?

   __Possible answer: Both survey part of a population.__

6. How is an unbiased sample different from a biased sample?

   __Possible answer: An unbiased sample represents the population;__
   __a biased sample does not.__

Puzzles, Twisters & Teasers
9-1 Meltdown!

Find and circle the words below in the word search (horizontally, vertically or diagonally). Find a word that answers the puzzle. Circle the word and write it on the line.

| population | sample | biased | random | systematic |
| stratified | method | average | exit | poll |

What do you call a snowman with a tan?

A ____PUDDLE____.

LESSON 9-2 Practice A
Organizing Data

1. Complete the line plot to organize the data of math quiz scores.

   Math Quiz Scores
   18 18 20 13 17 12 15 12
   17 19 17 18 18 20 11 19

   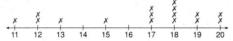

2. List the data values in the stem-and-leaf plot.

     1, 2, 5, 10, 15, 22,

   Key: 3 | 7 = 37   24, 26, 31, 37,

3. Use the given data to make a stem-and-leaf plot.

Maximum Speed of Animals (mph)			
pig (domestic)	11	grizzly bear	30
squirrel	12	rabbit	35
elephant	25	zebra	40
cat (domestic)	30	cape hunting dog	45

   1 | 1 2
   2 | 5
   3 | 0 0 5
   4 | 0 5
   Key: 4 | 5 = 45

4. Make a Venn diagram to show how many boys in an eighth-grade class had summer jobs.

Gender	M	M	F	M	F	F	M	F	F	M	M	M
Summer Job?	yes	no	yes	yes	yes	no	no	yes	yes	yes	no	yes

   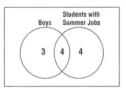

LESSON 9-2 Practice B
Organizing Data

1. Use a line plot to organize the data of the distances students travel to school.

   Distances Students Travel to School (mi)
   2 8 6 10 5 4 6 8 3 2
   11 5 1 3 6 5 7 5 2 4

2. List the data values in the stem-and-leaf plot.

   2 | 0 1 5 7
   3 | 2 2 9   20, 21, 25, 27, 32, 32, 39,
   4 | 5 6 7 9
   5 | 1 3   Key: 5 | 1 = 51   45, 46, 47, 49, 51, 53

3. Use the given data to make a back-to-back stem-and-leaf plot.

   NBA Midwest Division 2000–2001 Final Standings

NBA Team	Wins	Losses	NBA Team	Wins	Losses
San Antonio Spurs	58	24	Houston Rockets	45	37
Utah Jazz	53	29	Denver Nuggets	40	42
Dallas Mavericks	53	29	Vancouver Grizzlies	23	59
Minnesota Timberwolves	47	35			

Wins		Losses
3	2	4 9 9
	3	5 7
7 5 0	4	2
8 3 3	5	9
Key: 5	9 represents 59
3	5 represents 53

4. Make a Venn diagram to show how many girls in an eighth-grade class belonged to both a team and a club.

Team	yes	no	yes	no	yes	yes	yes	no	yes	no	no
Club	yes	yes	no	yes	yes	no	yes	yes	no	no	yes

   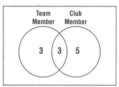

LESSON 9-2 Practice C
Organizing Data

1. Use a line plot to organize the data of dog temperatures measured at a veterinary's office one day.

   Dog Temperatures (F°)
   101.3 101.6 101.5 102.3 101.9 102.1
   101.5 102.2 102.0 101.8 102.1 101.8

2. List the data values in the stem-and-leaf plot.

   0 | 6
   1 | 1 5 6 8
   2 | 3   0.6, 1.1, 1.5, 1.6, 1.8, 2.3,
   3 | 0 7 9
   4 | 1 2 4   Key: 4 | 1 = 4.1   3.0, 3.7, 3.9, 4.1, 4.2, 4.4

3. Use the given data to make a back-to-back stem-and-leaf plot.

Animal	Endangered	Threatened
Mammals	63	9
Birds	78	14
Reptiles	14	22
Amphibians	10	8
Fishes	70	44
Clams	61	8
Snails	20	11
Insects	33	9
Arachnids	12	0

Endangered		Threatened
	0	0 8 8 9 9
4 2 0	1	1 4
0	2	2
3	3
	4	4
3 1	6
8 0	7
Key: 4	4 represents 44
0	7 represents 70

4. Make a Venn diagram to show how many students in an eighth-grade class have both a television and a computer in their bedroom.

Television	yes	yes	no	no	no	yes	yes	yes	yes	no	yes
Computer	yes	yes	no	yes	no	yes	no	yes	yes	no	yes

   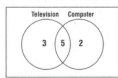

LESSON 9-2 Reteach
Organizing Data

Horizontal displays of numbers can be written in a compact form by eliminating repetition.

60 61 63 66 66 67 can be written 6 | 0 1 3 6 6 7

All the numbers start with 6, the **stem**.
Just write the second digit for each number, the **leaves**.

Display each set of numbers in compact form, using a stem and leaves.

1. 72 75 75 76 76 76 79

   7 | 2 5 5 6 6 6 9

2. 120 123 124 125 125 127

   12 | 0 3 4 5 5 7

Here are the scores on the last test in Ms. Kahn's math class.
76 84 88 93 97 65 100 86 91 97
93 79 81 99 92 78 78 79 87 100

To display these scores in a **stem-and-leaf plot**:

Use the given order to record scores.

6 | 5
7 | 6 9 8 8 9
8 | 4 8 6 1 7
9 | 3 7 1 7 3 9 2
10 | 0 0
Key: 7 | 6 represents 76.

Order each set of leaves.

6 | 5
7 | 6 8 8 9 9
8 | 1 4 6 7 8
9 | 1 2 3 3 7 7 9
10 | 0 0
Key: 7 | 6 represents 76.

Complete a stem-and-leaf plot for the data.

3. Daily High Temperatures
   46 52 48 47 56 59 61 50 37 35 34 37 44 49 43
   43 44 50 54 52 55 50 48 46 44 39 37 32 32 32

   3 | 7 5 4 7 9 7 2 2 2
   4 | 6 8 7 4 9 3 3 4 8 6 4
   5 | 2 6 9 0 0 0 2 3 0
   6 | 1
   Key: 5 | 2 represents 52.

   3 | 2 2 2 4 5 7 7 7 9
   4 | 3 3 4 4 4 6 6 7 8 8 9
   5 | 0 0 0 0 2 2 3 6 9
   6 | 1
   Key: 5 | 2 represents 52.

4. Create a stem-and-leaf plot for heights in inches of students in Mrs. Gray's class.
   40 48 49 53 60 62 48 62 53 54 60 65 63 49 55

   4 | 0 8 8 9 9
   5 | 3 3 4 5 5
   6 | 0 0 2 2 3 5   Key: 4 | 0 represents 40.

LESSON 9-2 Reteach
Organizing Data (continued)

Two sets of data can be compared by using a **back-to-back stem-and-leaf plot**.

This plot compares the number of games won to the number of games lost by each of the pennant winners in the American League East for the years 1995–2000.

The stems are in the center.
The left leaves are read in reverse.
The greatest number of games won was 114.
The greatest number of games lost was 74.

American League East Pennant Winners 1994–2000

Games Lost		Games Won
8	4	
8	5	
4 4	6	
4 0	7	
	8	6 7
	9	2 8 8
	10	
	11	4

Key: | 8 | 6 represents 86 games.
Key: 8 | 4 | represents 48 games.

Refer to the American League East back-to-back stem-and-leaf plot shown above.

5. The least number of games won was: _____86_____
6. The least number of games lost was: _____48_____

Using the given data set, complete to make a back-to-back stem-and-leaf plot.

7. **American League West Pennant Winners**

Year	Winner	Won	Lost
1994	Texas	52	62
1995	Seattle	79	66
1996	Texas	90	72
1997	Seattle	90	72
1998	Texas	88	74
1999	Texas	95	67
2000	Oakland	91	70

Games Lost		Games Won
	5	2
7 6 2	6	
4 2 2 0	7	9
	8	8
	9	0 0 1 5

Key: | 8 | 8 represents 88 games.
Key: 2 | 6 | represents 62 games.

Refer to the American League West back-to-back stem-and-leaf plot.

8. The difference between the greatest number of games won and the least number of games won was: _____95 − 52 = 43_____

9. The difference between the greatest number of games lost and the least number of games lost was: _____74 − 62 = 12_____

LESSON 9-2 Challenge
Pet Sets

A Venn diagram can show the relationships among three sets of data. Use the survey results shown in the table to complete the Venn diagram.

Number of Students in Mr. Phillips' Math Class Whose Families Have Pets

Dogs	Dogs and Cats	Dogs and Fish	Dogs, Cats, and Fish	Cats	Cats and Fish	Fish
13	7	2	2	12	3	6

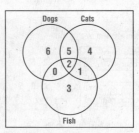

Use your Venn diagram to answer the questions.

1. How many of the families have dogs only? _____6_____
2. How many of the families have cats only? _____4_____
3. How many of the families have fish only? _____3_____
4. How many of the families have dogs and fish but no cats? _____0_____
5. How many of the families have cats and fish but no dogs? _____1_____

LESSON 9-2 Problem Solving
Organizing Data

A consumer survey gathered the following data about what teens do while on online.

1. Make a stem-and-leaf plot of the data.

Teens' Activities Online

7 | 3 3
8 | 2 6
9 | 5

Key: 7 | 3 means 73%

Teens' Activities Online	
Activity	Percent
E-mail	95
Use search engines	86
Instant Messaging	82
Visit music sites	73
Enter contests	73

The stem-and-leaf plot that shows the total number of medals won by different countries in the 2000 Summer Olympics. Choose the letter for the best answer.

2. List all the data values in the stem-and-leaf plot.
 A 2, 4, 5, 6, 7, 8, 9
 B 23, 25, 26, 28, 28, 29, 34, 38, 40, 57, 58, 59, 60, 70, 88, 97
 C 23, 25, 26, 28, 29, 34, 38, 57, 58, 59, 88, 97
 Ⓓ 23, 25, 26, 28, 28, 29, 34, 38, 57, 58, 59, 88, 97

2000 Olympic Medals

2 | 3 5 6 8 8 9
3 | 4 8
4 |
5 | 7 8 9
6 |
7 |
8 | 8
9 | 7

3. What is the least number of medals won by a country represented in the stem-and-leaf plot?
 F 3
 G 4
 Ⓗ 23
 J 97

4. What is the greatest number of medals won by a country represented in the stem-and-leaf plot?
 A 9
 B 70
 C 79
 Ⓓ 97

LESSON 9-2 Reading Strategies
Use a Graphic Organizer

Below is a list of students' scores for a science test.

75 84 90 68 73 83 95 85 89 97 77 72 83

A stem-and-leaf plot is one way to organize the test scores.

stem | leaf
6 | 5 8
7 | 2 3 5 7
8 | 3 3 4 5 9
9 | 0 5 7

Use your stem-and-leaf plot to answer the questions.

1. What is the stem of the score 83? _____8_____
2. What is the leaf of the score 90? _____0_____
3. What is the highest score? _____97_____
4. What is the lowest score? _____65_____
5. What is the difference between the highest and lowest scores? _____32_____

Below is a list of students' scores for a social studies test.

69 84 75 99 65 70 81 87 73 81 92 95 71

6. Make a stem-and-leaf plot of the scores.

stem | leaf
6 | 5 9
7 | 0 1 3 5
8 | 1 1 4 7
9 | 2 5 9

7. Can you find the lowest and highest scores more easily by looking at the list or at the stem-and-leaf plot?

_____stem-and-leaf plot_____

LESSON 9-2 Puzzles, Twisters & Teasers
A Bright Idea!

List the values in the stem-and-leaf plot from the top line down. Each answer has a corresponding letter. Use the letters to solve the riddle.

```
3 | 2 5
4 | 0 3 7
5 | 1 4 8
6 | 1 6
```

T	32
H	35
W	40
E	43
G	47
R	51
I	54
S	58
B	61
O	66

Why did the teacher wear sunglasses in class?

Because her students W E R E   S O   B R I G H T
                      40 43 51 43   58 66   61 51 54 47 35 32

LESSON 9-3 Practice A
Measures of Central Tendency

Find the mean, median, mode, and range of each set of numbers.

1. 4, 2, 6, 3, 8, 6, 6
   mean: __5__ mode: __6__
   median: __6__ range: __6__

2. 2, 8, 6, 9, 8, 7, 9, 8
   mean: __7.125__ mode: __8__
   median: __8__ range: __7__

3. 12, 9, 14, 22, 3, 11, 14, 15
   mean: __12.5__ mode: __14__
   median: __13__ range: __19__

4. 89, 45, 68, 94, 70, 94, 86
   mean: __78__ mode: __94__
   median: __86__ range: __49__

Determine and find the most appropriate measure of central tendency or range for each situation. Refer to the table.

Waterfall Heights (ft)	
Feather, CA	640
Bridalveil, CA	620
Ribbon, NV	1,612
Seven, CO	300
Akaka, HI	442
Shoshone, ID	212
Taughannock, NY	215
Multnomah, OR	620

5. What number best describes the middle of the waterfall heights?
   median; 531 ft

6. What number appears most often in the waterfall heights?
   mode; 620 ft

7. Which measure of central tendency is best to describe the waterfall heights? Explain your reasoning.
   Possible answer: The median is best because it eliminates the influence of the outlier (1,612 ft for Ribbon Falls).

8. An official for the department of transportation counted the number of vehicles that passed through a busy intersection. He counted for 10 consecutive minutes and recorded the number of vehicles for each minute: 18, 41, 25, 9, 22, 36, 24, 13, 25, and 28. What number best describes the middle of the data?
   mean = 24.1

LESSON 9-3 Practice B
Measures of Central Tendency

Find the mean, median, mode, and range of each data set.

1. 7, 7, 4, 9, 6, 4, 5, 8, 4
   mean: __6__
   median: __6__
   mode: __4__
   range: __5__

2. 1.2, 5.8, 3.7, 9.7, 5.5, 0.3, 8.1
   mean: __4.9__
   median: __5.5__
   mode: __none__
   range: __9.4__

3. 31, 28, 31, 30, 31, 30, 31, 31, 30, 31, 30, 31
   mean: __30.416__
   median: __31__
   mode: __31__
   range: __3__

4. 65, 46, 78, 3, 87, 12, 99, 38, 71, 38
   mean: __53.7__
   median: __55.5__
   mode: __38__
   range: __96__

Determine and find the most appropriate measure of central tendency or range for each situation. Refer to the table at the right for Exercises 5–7.

Some Major Earthquakes in United States History

Year	Location	Magnitude
1812	Missouri	7.9
1872	California	7.8
1906	California	7.7
1957	Alaska	8.8
1964	Alaska	9.2
1965	Alaska	8.7
1983	Idaho	7.3
1986	Alaska	8.0
1987	Alaska	7.9
1992	California	7.6

5. Which measure best describes the middle of the data?
   ~~mean; 8.09~~ median; 7.9

6. Which earthquake magnitude occurred most frequently?
   mode; 7.9

7. How spread out are the data?
   range; 1.9

8. Nicole purchased gasoline 8 times in the last two months. The prices that she paid per gallon each time were $2.19, $2.14, $2.28, $2.09, $2.01, $1.99, $2.19, and $2.39. Which measure makes the prices appear lowest?
   mean; $2.16

LESSON 9-3 Practice C
Measures of Central Tendency

Find the mean, median, mode, and range of each data set.

1. 94, 90, 88, 66, 94, 81, 102, 108, 88
   mean: __90.1̄__
   median: __90__
   mode: __88 and 94__
   range: __42__

2. 16.58, 9.99, 24.30, 48.85, 25.09, 9.71
   mean: __22.42__
   median: __20.44__
   mode: __none__
   range: __39.14__

3. 173, 160, 232, 148, 162, 160, 265, 182, 98, 147, 162, 205, 162, 169
   mean: __173.21__
   median: __162__
   mode: __162__
   range: __167__

4. 2386, 3154, 2873, 4256, 3184, 2389, 3141, 2452, 3000, 2584, 4189
   mean: __3055.27̄__
   median: __3000__
   mode: __none__
   range: __1870__

Determine and find the most appropriate measure of central tendency or range for each situation. For Exercises 5–7, refer to the table at the right.

Difference Between the Highest and Lowest Points of Each Continent

Continent	Difference (Feet)
North America	20,602
South America	22,965
Europe	18,602
Asia	30,347
Africa	19,852
Australia & Oceania	7,362
Antarctica	25,191

5. Which measure best describes the middle of the data?
   median; 20, 602 ft

6. Which difference in altitude occurs most frequently?
   none; There is no mode.

7. How spread out are the data?
   range; 22,985 feet

8. As of 2001, the estimated population, in millions, of each continent is as follows: North America, 476; South America, 343; Europe, 727; Asia, 3641; Africa, 778; Australia & Oceania, 30; and Antarctica, 0. Which measure makes the population appear greatest?
   mean; 856.43 million

Reteach
9-3 Measures of Central Tendency

The **mode** of a data set is the value (or values) that occur(s) most often.

2, 4, 10, **3**, 6, **3**, 7
The value 3 occurs most often. So, 3 is the mode.

The **median** of a data set is the middle value—after the values have been ordered.

2, 4, 10, 3, 6, 3, 7 → 2, 3, 3, **4**, 6, 7, 10
The middle value is 4. So, 4 is the median.

The **mean** of a data set is the average value. Add the values and divide the sum by the number of values in the set.

2, 4, 10, 3, 6, 3, 7 → $\frac{2+4+10+3+6+3+7}{7} = \frac{35}{7}$, or 5
So, the mean is 5.

The **range** of a data set is the difference between the greatest value and the least value.

2, 4, **10**, 3, 6, 3, 7 → 10 − 2 = 8
So, the range is 8.

Determine and find the most appropriate measure of central tendency or range for each situation. Refer to the table at the right.

Smith Family Reunion 2006

Family Member	Age
Aunt Beth	36
Uncle Steve	40
Louise	9
Travis	6
Grandma	62
Grandpa	62
Mom	44
Dad	43
Me	13

1. Which age was most frequent at the reunion?
   mode; 62 years

2. Which age was the average at the reunion?
   mean; 35 years

3. How spread out are the ages?
   range; 56 years

4. What was the middle age?
   median; 40 years

Challenge
9-3 The Groupie Effect

A class of 29 students reported the number of books read so far this school year.

Number of Books Read
5, 5, 6, 3, 6, 3, 2, 7, 5, 3,
7, 4, 2, 5, 6, 7, 6, 4, 1, 4,
9, 5, 6, 7, 7, 6, 6, 7, 5

This frequency table shows the same data.

Value	1	2	3	4	5	6	7	8	9
Frequency	1	2	3	3	6	7	6	0	1

1. Explain how to find the mode using the ungrouped data. What is the mode?
   Look for the score with the highest frequency; mode = 6

2. Explain how to find the mode using the frequency table. Verify that you obtain the same result as before.
   Look for the interval with the highest frequency; mode = 6

3. Explain how to find the median using the ungrouped data. What is the median?
   Arrange the data in order. The median is the middle, or 15th score. median = 5

4. Explain how to find the median using the frequency table. Verify that you obtain the same result as before. Which method do you prefer? Why?
   From either end of the frequency row of the table, add frequencies until you get to 15. median = 5. Preferences vary.

5. Explain how to find the mean using the ungrouped data. What is the mean?
   Add the 29 scores and divide the sum by 29. mean = 5.14

6. Explain how to find the mean using the frequency table. Verify that you obtain the same result as before. Which method do you prefer? Why?
   Multiply each interval by its frequency. Add the products. Divide the sum by 29. mean = 5.14 Preferences will vary.

Problem Solving
9-3 Measures of Central Tendency

Use the data to find each answer.

World's Busiest Airports

Airport	Total Passengers (in millions)
Atlanta, Hartsfield	80.2
Chicago, O'Hare	72.1
Los Angeles	68.5
London, Heathrow	64.6
Dallas/Ft. Worth	60.7

1. Find the average number of passengers in the world's five busiest airports.
   69.22 million

2. Find the median number of passengers in the world's five busiest airports.
   68.5 million

3. Find the mode of the airport data.
   There is no mode.

4. Find the range of the airport data.
   19.5 million

Choose the letter for the best answer.

World Motor Vehicle Production (in thousands) 1998–1999

Country	1998	1999
United States	12,047	13,063
Canada	2,568	3,026
Europe	16,332	16,546
Japan	10,050	9904

5. What was the mean production of motor vehicles in 1998?
   A 8,651,500 vehicles
   B 10,249,250 vehicles
   C 11,264,250 vehicles
   D 12,000,000 vehicles

6. What was the range of production in 1999?
   F 9,800,000 vehicles
   G 11,480,000 vehicles
   H 12,520,000 vehicles
   J 13,520,000 vehicles

7. What was the median number of vehicles produced in 1999?
   A 3,026,000 vehicles
   B 3,069,000 vehicles
   C 11,483,500 vehicles
   D 13,063,000 vehicles

8. Which value is largest?
   F Mean of 1998 data
   G Mean of 1999 data
   H Median of 1998 data
   J Median of 1999 data

Reading Strategies
9-3 Vocabulary Development

The **mean**, the **median**, and the **mode** are measures that tell about the middle part of a set of data. This chart will help you learn about each of them.

Mean (average): The sum of all values divided by total number of values

Median: The middle number, or the average of the two middle numbers

↓ ↓

Measures of the Middle Part of a Set of Data

↓

Mode: The value or values that occur most often

Use the chart to answer each question.

1. What is the median?
   It is the middle number in a set of data, or the average of the two middle numbers.

2. What is the mean?
   It is the sum of all values divided by the total number of values.

3. What is a mode?
   It is the value or values that occur most often.

The following list shows the number of goals scored each month by a hockey team.
4 4 7 6 8 3 5

Answer each question.

4. What is the mode of the data?
   4 goals

5. What is the median of the data?
   5 goals

LESSON 9-3 Puzzles, Twisters & Teasers
Math a la Mode!

Find the mean, median, or mode for each data set. Each answer has a corresponding letter. Use the letters to solve the riddle.

Find the mean.
1. 20, 17, 42, 26, 27, 12, 31   25   E
2. 15, 10, 12, 10, 13, 13, 10, 3   10.75 S
3. 8, 5, 3, 75, 7, 3, 4, 7, 9, 2, 8, 5, 7   11   T
4. 4.0, 3.3, 5.6, 4.6, 3.3, 5.6   4.4 P

Find the median.
5. 35, 21, 34, 44, 36, 42, 29   35   A
6. 23, 13, 45, 56, 72, 44, 89, 92, 67   56   L
7. 5, 2, 12, 7, 13, 9, 8   8   O
8. 4.3, 1.3, 4.5, 8.6, 9, 3, 2.1   4.3 H

Find the mode.
9. 92, 88, 84, 86, 88   88   C
10. 6, 8, 6, 7, 9, 2, 4, 22   6   I
11. 7, 5, 4, 6, 8, 3, 5, 2, 5   5   T
12. 2.0, 4.5, 6.2, 3.2, 4.5, 6.5   4.5 N

Why did the elephant sit on the marshmallow?

It didn't want to

S P L A S H
10.75 4.4 56 35 10.75 4.3

I N T O
6 4.5 5 8

T H E
11 4.3 25

H O T
4.3 8 5

C H O C O L A T E
88 4.3 8 88 8 56 35 11 25

LESSON 9-4 Practice A
Variability

Find the least value, greatest value, and median for each data set.

1. 6, 9, 3, 7, 8, 7, 5
   least value: 3
   greatest value: 9
   median: 7

2. 12, 8, 24, 19, 15, 20, 13
   least value: 8
   greatest value: 24
   median: 15

Find the given values for each data set. Then use the values to make a box-and-whisker plot.

3. 27, 33, 28, 26, 34, 40, 21
   least value: 21
   greatest value: 40
   median: 28
   first quartile: 26
   third quartile: 34

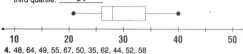

4. 48, 64, 49, 55, 67, 50, 35, 62, 44, 52, 58
   least value: 35
   greatest value: 67
   median: 52
   first quartile: 48
   third quartile: 62

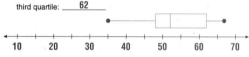

LESSON 9-4 Practice B
Variability

Find the first and third quartiles for each data set.

1. 37, 48, 56, 35, 53, 41, 50
   first quartile: 37
   third quartile: 53

2. 18, 20, 34, 33, 16, 44, 42, 27
   first quartile: 19
   third quartile: 38

Use the given data to make a box-and-whisker plot.

3. 55, 46, 70, 36, 43, 45, 52, 61

4. 23, 34, 31, 16, 38, 42, 45, 30, 28, 25, 19, 32, 53

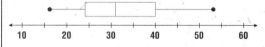

Use the box-and-whisker plots to compare the data sets.

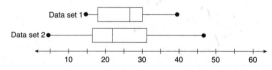

5. Compare the medians and ranges.

The median of data set 1 is greater than the median of data set 2. The range of data set 2 is greater than the range of data set 1.

6. Compare the ranges of the middle half of the data for each set.

The range of the middle half of the data is greater in data set 2.

LESSON 9-4 Practice C
Variability

Use the given data to make a box-and-whisker plot.

1. 76, 53, 55, 64, 43, 67, 73, 82, 71, 49, 58, 64

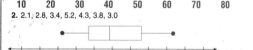

2. 2.1, 2.8, 3.4, 5.2, 4.3, 3.8, 3.0

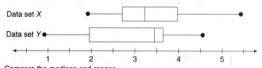

Use the box-and-whisker plots to compare the data sets.

3. Compare the medians and ranges.

The median of data set Y is greater than the median of data set X. The ranges of data set X and data set Y are almost equal.

4. Compare the ranges of the middle half of the data for each set.

The range of the middle half of the data is greater in data set Y.

Match each set of data with a box-and-whisker plot.

5. range: 32; first quartile: 17; third quartile: 34   plot c
6. range: 22; first quartile: 17; third quartile: 29   plot a
7. range: 23; first quartile: 12; third quartile: 25   plot b

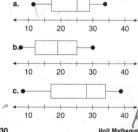

LESSON 9-4 Reteach
Variability

Quartiles divide a data set into four equal parts.

1 1 3 3 3 4 4 6 6 6 6	6	7 7 7 7 7 8 9 9 9 9 9

The **1st quartile** is the median of the lower half.
The **median**, or **2nd quartile**, divides the data into two halves.
The **3rd quartile** is the median of the upper half.

For each data set, circle and label the median *M*. Circle and label the first quartile Q_1. Circle and label the third quartile Q_3.

1. 3 ③ 3 ④ 5 ⑥ 7
      Q_1 M Q_3

2. 3 3 ③ 4 ④ 5 ⑤ 6 7 ⑨ 10 10
        Q_1 M Q_3

A **box-and-whisker** plot displays the quartile values as well as the lowest and highest numbers of a data set. The sides of the box are the first and third quartiles; the median is inside the box. The whiskers connect the box to the lowest and highest values.

This plot displays the values for the data set at the top of this page.

lowest value = 1
1st quartile = 4
median = 6
3rd quartile = 8
highest value = 9

Complete to make a box-and-whisker plot for the data set 45, 47, 47, 48, 48, 49, 53.

3. First, calculate three significant values for the data set.

   median = __48__   1st quartile = __47__   3rd quartile = __49__

LESSON 9-4 Challenge
What's Normal?

Standard deviation (symbol σ, *sigma*) is a measure of variability that tells how far data are spread out from the mean of a data set.

In many situations, such as scores on the SAT or other standardized tests, the data cluster around the mean in such a way that if they are graphed to show the frequency of measures, the graph appears as a **bell-shaped curve**, also called the **normal curve**.

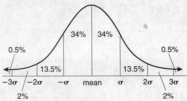

If the mean math score for males on the 2001 SAT I was 533 and the standard deviation was 115, determine the scores achieved by about 68% of the male participants.

According to the normal curve, 68% of scores fall between $-\sigma$ and σ.
mean $- \sigma = 533 - 115 = 418$ mean $+ \sigma = 533 + 115 = 648$
So, the scores for about 68% of the males fell between 418 and 648.

Assume a normal distribution for each situation.

1. A survey of 16-year-olds showed that they watched an average (mean) of 9.4 hours of TV per week, with a standard deviation of 1.2 hours. Determine how many hours of TV were watched by about:

   a. 68% of the participants. __between 8.2 and 10.6 hours__

   b. 95% of the participants. __between 7 and 11.8 hours__

2. On a certain standardized test, the mean score was 50 and the standard deviation 3. About what percent of the participants scored:

   a. between 50 and 56? __47.5%__

   b. 44 and 47? __13.5%__

LESSON 9-4 Problem Solving
Variability

Write the correct answer.

1. Find the median of the data.
   __14.5 points__

2. Find the first and third quartiles of the data.
   $Q_1 = 10$, $Q_3 = 23$

3. Make a box-and-whisker plot of the data.

Super Bowl Point Differences

Year	Point Difference
2001	27
2000	⑦
1999	15
1998	7
1997	14
1996	10
1995	23
1994	17
1993	㉟
1992	13

The box-and-whisker plots compare the highest recorded Fahrenheit temperatures on the seven continents with the lowest recorded temperatures. Choose the letter for the best answer.

4. Which statement is true?
   A The median of the high temperatures is less than the median of the low temperatures.
   B The range of low temperatures is greater than the range of high temperatures.
   C The range of the middle half of the data is greater for the high temperatures.
   D The median of the high temperatures is 49°F.

5. What is the median of the high temperatures?
   F 128°F   H −67°F
   G 120°F   J −90°F

6. What is the range of the low temperatures?
   A 77°F   **C** 120°F
   B 79°F   D 129°F

LESSON 9-4 Reading Strategies
Analyze Information

One way to organize data is to divide it into four equal parts called **quartiles**.

Here are the ages of students in the school chorus:
12 12 12 13 13 13 14 14 15 15 15

To organize this data into **quartiles**:

- First, find the **median**, the value that divides the set of data in half.

  Lower half   Upper half
  12 12 12 13 13 ⑬ 14 14 15 15 15
                     Median

1. What value divides the data in half?
   __13__

- The first and second quartiles are divided by the middle value of data *below* the median.

  1st Quartile   2nd Quartile   3rd Quartile   4th Quartile
  12 12 ⑫ 13 13 ⑬ 14 14 ⑮ 15 15

2. What is the middle value below the median?
   __12__

- The third and fourth quartiles are divided by the middle value of the data *above* the median.

3. What is the middle value above the median?
   __15__

4. When a group of data is organized into four equal parts, what is each part called?
   __a quartile__

Circle the quartile in which each age is located.

5. 12   (1st quartile)   2nd quartile   3rd quartile   4th quartile
6. 14   1st quartile   2nd quartile   (3rd quartile)   4th quartile

LESSON 9-4 Puzzles, Twisters & Teasers
What Are Your Values?

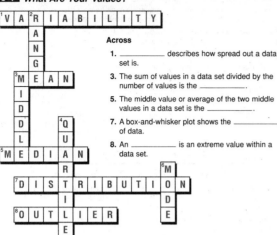

Across

1. _____ describes how spread out a data set is.
3. The sum of values in a data set divided by the number of values is the _____.
5. The middle value or average of the two middle values in a data set is the _____.
7. A box-and-whisker plot shows the _____ of data.
8. An _____ is an extreme value within a data set.

Down

2. The _____ of a data set is the largest value minus the smallest value.
3. A measure of central tendency describes the _____ of a data set.
4. A _____ divides a data set into four equal parts.
6. The value or values that occur most often in a data set is the _____.

Crossword answers: 1-VARIABILITY, 2-RANGE, 3-MEAN, 3d-MIDDLE, 4-QUARTILE, 5-MEDIAN, 6-MODE, 7-DISTRIBUTION, 8-OUTLIER

LESSON 9-5 Practice A
Displaying Data

1. Use the data to complete the double-bar graph.

Softball Scores	1	2	3	4	5	6	7	8
Team A	3	2	0	2	1	4	2	1
Team B	1	4	3	0	3	1	2	1

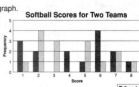

2. Use the data to make a histogram with intervals of 10.

Average High Temperatures in April in Tourist Cities			
Acapulco, Mexico	87	Montreal, Canada	51
Athens, Greece	67	Nassau, Bahamas	81
Dublin, Ireland	54	Paris, France	60
Hong Kong, China	79	Rome, Italy	68
London, U.K.	56	Sydney, Australia	73
Madrid, Spain	63	Toronto, Canada	51

3. Make a double-line graph of the given data. Use the graph to estimate the heights of Dean and Susan when they were 9 years old.

Age	Dean's Height	Susan's Height
2	35 in.	30 in.
4	41 in.	37 in.
6	46 in.	44 in.
8	50 in.	51 in.
10	57 in.	58 in.
12	60 in.	65 in.

At 9 years old, Dean was approximately __53.5 in.__ tall, and Susan was approximately __54.5 in.__ tall.

LESSON 9-5 Practice B
Displaying Data

1. Make a double-bar graph.

Daily Hours Worked	6	7	8	9	10	11	12
Crew A	4	3	6	1	3	1	2
Crew B	5	5	4	3	2	0	1

2. Use the data to make a histogram with intervals of 5.

Weekly Allowance of 20 Students			
$5	$15	$2	$10
$12	$12	$10	$15
$10	$5	$6	$4
$8	$7	$20	$7
$5	$4	$5	$9

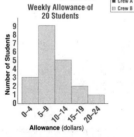

3. Make a double-line graph of the given data. Use the graph to estimate the number of radio stations and cable TV systems in 2002.

Commercial Media in the United States		
Year	Radio Stations	Cable TV Systems
1997	10,207	10,950
1999	10,444	10,700
2001	10,516	9,924
2003	10,605	9,339

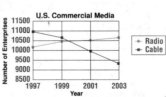

Possible answer: From the graph, I would estimate there were about 10,550 radio stations and about 9,700 cable TV systems in 2002.

LESSON 9-5 Practice C
Displaying Data

1. Maggie found hotels in New York City offering the following rates per night. Make a histogram of the rates with intervals of $50.

$175 $134 $119 $275 $318
$155 $99 $178 $202 $199
$280 $160 $108 $137 $151
$221 $299 $97 $148 $283

2. Make a double-line graph of the given data. Use the graph to answer questions 3 and 4.

Railroad Ridership (in millions)										
	1991	1992	1993	1994	1995	1996	1997	1998	1999	2000
Amtrack	22.0	21.3	22.1	21.2	20.7	19.7	20.2	21.1	21.5	22.5
Commuter	18.1	20.3	32.9	39.5	42.2	45.9	48.5	54.0	58.3	61.6

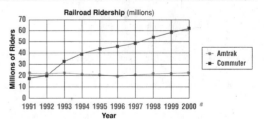

3. Do you predict the number of riders increased or decreased in 2001?

__increase__

4. Between which years did the number of commuter train riders increase the most?

__between 1992 and 1993__

LESSON 9-5 Reteach
Displaying Data

A double-bar graph is used to compare quantities measured in the same unit.

The bars may be vertical.

The bars may be horizontal.

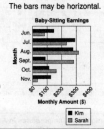

For Exercises 1-3, refer to the table below.

League Champion Batting Averages

Year	1997	1998	1999	2000	2001
National League	0.372	0.363	0.379	0.372	0.350
American League	0.347	0.339	0.357	0.372	0.350

1. If the horizontal axis will show the years and the vertical axis will show the batting averages, will you use vertical bars or horizontal bars for the graph?

   vertical

2. To avoid decimals for the vertical scale, include *(in thousandths)* as part of the label. Then, what are appropriate values for the vertical scale?

   Possible answer: 340–380 in intervals of 10

3. Draw a double-bar graph for the data.

LESSON 9-5 Reteach
Displaying Data (continued)

A **histogram** represents intervals of grouped data as bars with no space between. The data may be organized in a **frequency table**.

The lengths of selections most requested today by listeners of station WXYZ are shown in a frequency table and a histogram.

Frequency Table

Selection Length (in seconds)	Frequency (number of requests)
60–119	2
120–179	5
180–239	12
240–299	14
300–359	7
360–419	2

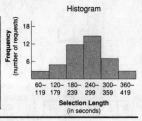

Complete to make a frequency table and histogram for this data set.

The heights, in inches, of 25 students in a high school class are 70, 72, 59, 56, 55, 60, 48, 72, 59, 48, 66, 72, 58, 60, 60, 50, 68, 72, 68, 62, 72, 58, 60, 68, 59.

4. Complete this frequency table for the data.

Interval	Tallies	Frequency
70–74	||||		6
65–69	||||	4
60–64	||||	5
55–59	|||| ||	7
50–54			1
45–49	||	2

5. Use the frequency table to make a histogram.

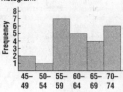

6. Discuss how the histogram for this data would change if the intervals were 40–49, 50–59, 60–69, 70–79.

There would be 4 bars instead of 6. The new bars would have higher frequencies. The frequency of most bars would equal the sum of the frequencies of two bars in the graph above.

LESSON 9-5 Challenge
Double the Fun

A **double-bar graph** is used to compare between data groups and within data groups.

1. In which quarter did the company experience its greatest profit? Explain.

   4th quarter;
   profit = income − expenses

2. Did the company experience a profit or loss for the year? Estimate the amount.

   profit of about $50,000

3. Make a double-bar graph to display this data.

Baseball Home-Run Leaders, years 1997–2001

	1997	1998	1999	2000	2001
American League	56	56	48	47	52
National League	49	70	65	50	73

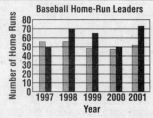

4. In which year was the difference between home-run leaders greatest?

   2001

5. Compare the home-run leader data for the American League and the National League.

   Possible answer: In most years, the leader for the National League hit more runs than the leader for the American League.

LESSON 9-5 Problem Solving
Displaying Data

Make the indicated graph.

1. Make a double-bar graph of the homework data.

Hours of Daily Homework	1	2	3	4	5
Boys	12	5	2	1	0
Girls	4	6	5	3	2

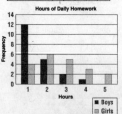

2. The annual hourly delay per driver in the 20 U.S. cities with the most traffic are as follows: 56, 42, 53, 46, 34, 37, 42, 34, 53, 21, 45, 50, 34, 42, 41, 38, 42, 34, 38, 31. Make a histogram with intervals of 5 hours.

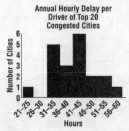

For 3–5, refer to the double-line graph. Circle the letter of the correct answer.

3. Estimate the value of U.S. agricultural exports in 1998.
   A $62 billion
   B $59 billion
   Ⓒ $52 billion
   D Cannot be determined

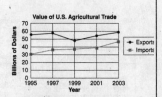

4. Estimate the value of U.S. agricultural imports in 2000.
   Ⓕ $39 billion
   G $31 billion
   H $29 billion
   J $21 billion

5. Estimate the difference between agricultural exports and imports in 1995.
   A $16 billion
   B $21 billion
   Ⓒ $26 billion
   D Cannot be determined

LESSON 9-5 Reading Strategies
Use Graphic Aids

A **line graph** shows changes in data over time.
- The temperature changes from morning to night.
- Your age changes from year to year.

This double-line graph shows money earned from ice cream and hot dogs sold during a class picnic.

Use the line graph to answer each question.

1. What information is found along the left side of the graph?

   total sales in dollars

2. The values on the left side of the graph increase by what amount?

   by $20

3. What information is shown along the bottom of the graph?

   time of day

Each point on the graph shows total sales up to that time.

4. How much money was made from ice cream and hot dog sales by 11:00 A.M.?

   $60

5. During which time period was the least amount of ice cream sold?

   from 12:00 noon until 1:00 P.M.

LESSON 9-5 Puzzles, Twisters & Teasers
Cold Facts!

Find and circle the words from the list in the word search (horizontally, vertically or diagonally). Find a word that answers the riddle. Circle it and write it on the line.

bar   graph   frequency   table   histogram
line   display organize   data   random

```
Q T Y H B G U J N M I K O P
W I N D O W S D A T A V H A
R F T R R U I I N A N J I S
A P S D G F T S E B A R S D
N T G B A R D P I U Y T T F
D Z X C N W S L I N E A O B
O N M K I G R A P H J B G N
M L O I Z Q W Y V F T L R A
V G F R E Q U E N C Y E A Y
Q A Z X S W E D C V F R M X
```

Why did the computer catch a cold?

It forgot to close its _____WINDOWS_____

LESSON 9-6 Practice A
Misleading Graphs and Statistics

Explain why each graph is misleading.

1. The Price of a Pound of Apples in Selected Cities in September 2000

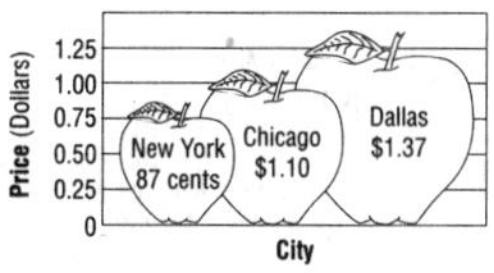

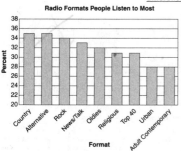

   Possible answers:

   The heights of the apples are used to represent prices. However, the areas of the apples distort the comparison. The area of the Dallas apple is about 3 times the area of the New York apple, but the price difference is only about 1.5 times.

2. Radio Formats People Listen to Most

   [bar graph showing Country ~35, Alternative ~35, Rock ~33, News/Talk ~32, Oldies ~32, Religious ~30, Top 40 ~31, Urban ~28, Adult Contemporary ~28; y-axis starts at 20]

   Because the scale does not start at 0, the bars for Country and Alternative are almost 100% higher than the bars for Adult Contemporary and Urban. In fact, they are only 25% more popular radio formats.

Explain why the statistic is misleading.

3. A juice company surveyed 4 people about which juice they preferred. Three of the people preferred the company's juice over the competition's. The company published that 3 times more people preferred their juice.

   The sample size is too small.

LESSON 9-6 Practice B
Misleading Graphs and Statistics

Explain why each graph is misleading.

1. On the Road
   Number of Trucks that Travel City Roads

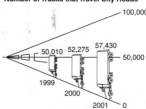

   1999: 50,010   2000: 52,275   2001: 57,430

   Possible answers:

   The heights of the truck bars vary greatly because the graph is not equally proportioned. The 2001 bar is about 3 times taller than the 1999 bar. In fact, the data for the 3 years is close together.

2. Federal Minimum Wage Rates Since 1980

   1980: $3.10   1981: $3.35   1990: $3.80   1991: $4.25   1996: $4.75   1997: $5.15

   Because the scale does not start at 0, the bar for 1997 is more than 21 times higher than the bar for 1980. In fact, the 1997 minimum wage is only about 1.7 times greater than the 1980 minimum wage.

Explain why the statistic is misleading.

3. A chewing gum company advertises that the flavor of its new chewing gum lasts for an average of 55 minutes based on the following durations reported by customers: 12 min, 33 min, 5 min, 200 min, and 25 min.

   Although 55 min is the average, only one person reported that the flavor lasted longer than 33 min. It is not likely that the flavor will last for 55 min.

Copyright © by Holt, Rinehart and Winston.

79

Holt Mathematics

Practice C
9-6 Misleading Graphs and Statistics

Explain why each graph is misleading.

1. Number of Units Sold   Possible answers are given:

- Portable CD Player
- DVD Player
- Big Screen TV

Different-sized icons represent one unit. The number of big screen TVs sold is less than one-third of the number of portable CD players sold, but it looks like more big screen TVs were sold.

2.

Showing only part of a graph distorts the data. The earnings for the summer quarter alone look like they are increasing rapidly. In fact, when compared with the earnings for the entire year, the summer quarter earnings look somewhat flat.

Explain why the statistic is misleading.

3. A hotel advertises that its average monthly temperature is 78°F. The average temperature for each month is 52°F, 57°F, 59°F, 90°F, 92°F, 98°F, 104°F, 95°F, 94°F, 93°F, 53°F, and 49°F.

Although 78°F is the average, no one month has an average temperature very close to 78°F. Each month is either considerably colder or hotter than 78°F.

Reteach
9-6 Misleading Graphs and Statistics

The most appropriate measure of central tendency should be used.
The scale of graphs should begin at 0, or use a broken scale.

Complete to tell why the situation is misleading.

When Harold got 70 on his math test, he told his family that the mode for the test was 70. He said, "More students got 70 than any other grade." The actual grades, which he did not tell his family, were 70, 70, 70, 81, 85, 86, 90, 94, 95, 96, 97, 100.

1. How did Harold mislead his family?
   He did not say that the mode of 70 was the lowest grade achieved.

2. What was the mean of the class? the median?
   about 86; 88

3. What would have been a more accurate way for Harold to tell his family how his grade compared with the other grades?
   Tell the mean or the median.

Complete to tell why the situation is misleading.

The graph shows the salaries at a small company.

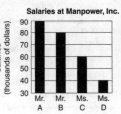

4. Use the heights of the bars to estimate the ratio of the salaries of Ms. C to Ms. D.
   3 to 1

5. According to the scale, what is Ms. C's salary? Ms. D's salary?
   $60,000; $40,000

6. What is the actual ratio of their salaries?
   3 to 2

7. Why is the graph misleading?
   The scale does not start at 0.

Challenge
9-6 To Tell the Truth...

For each situation, a statement based on the data is made. The statement is misleading. Explain why. **Possible answers:**

1. Ten students took the SAT I in their junior year. Then all ten took a course to prepare for the test the following year. Their average score on the math section increased from 525 to 565. The company offering the course stated in their new ads, "Students who took our course added 40 points to their math SAT."

   Scores did not necessarily increase as a result of the course. Students were a year older and had studied an additional year of math.

2. "Don't worry, Jack. You can dive here. The average depth of the pool is 8 ft."

   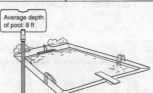

   Jack needs to be concerned about the depth of the pool where he dives, not the average depth.

3. Bill's test scores: 50, 52, 55, 63, 71, 98 Maria's test scores: 50, 85, 87, 92, 93, 98 "Both students' scores range from 50 to 98."

   The statement implies their scores are similar. But, Bill's average is much lower than Maria's.

4. An ad says "Save 50% on all regular price clothing! Take 20% off of all regular price clothing! Then take 30% of the newly reduced price for a total savings of 50%!

   Take 20% off a $100-item; result is $80. Then take 30% off the $80; result is $56, which is not half of the original $100.

5. A reporter asked twelve people if they went to the movies last week. Of the five who answered yes, he asked how many times they had gone to the movies. The responses were: 1, 1, 2, 2, and 4. The reporter then wrote "The average person attended two movies last week."

   The responses of seven of the twelve people interviewed are ignored.

Problem Solving
9-6 Misleading Graphs and Statistics

Explain why each statistics is misleading.   **Possible answers:**

1. A poll taken at a college says that 38% of students like pizza the best, 32% like hamburgers the best, and 30% like spaghetti the best. They conclude that most of the students at the college like pizza the best.

   62% of the students did not like pizza the best.

2. The National Safety Council of Ireland found that young men were responsible in 57% of automobile accidents they were involved in. The NSC Web site made this claim: "Young men are responsible for over half of all road accidents."

   Young men are not involved in all road accidents.

3. Explain why the Centers for Disease Control (CDC) has been highly criticized for the graph below.

   It implies that most Americans are killed by smoking.

   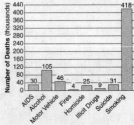

Choose the letter for the best answer.

4. Which statement is a misleading statistic for the data in the table?

Student	Test Grade
A	85%
B	92%
C	88%
D	10%
E	80%

   A The median score was 85%.
   B Most students scored an 80% or above.
   C The average test score was 71%.
   D The range of test scores was 82.

5. A sno-cone store claims, "Our sales have tripled." Sno-cone sales from March to May were 50 and sales from June to August were 150. Why is this misleading?

   A Sample size is too small.
   B During the summer, sales should be higher.
   C Should use the median not mean.
   D The statement isn't misleading.

LESSON 9-6 Reading Strategies
Compare and Contrast

Graph A and Graph B are bar graphs of the same data. A graph can be misleading if the numbers on the left side of the graph do not start at 0.

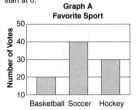
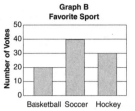

Answer each question.

1. Compare the titles of Graph A and Graph B.

   They are the same.

2. By how many does the number of votes increase on both graphs?

   by 10

3. Compare the first number shown along the left side of Graph A and the first number shown along the left side of Graph B.

   The first number along the left side of Graph A is 10 and the first number shown along the left side of Graph B is 0.

4. Compare the bars on Graph A to Graph B.

   The bars on Graph B are taller than the bars on Graph A.

5. Which graph is misleading? Why?

   Graph A; It looks like 3 times as many people prefer soccer to basketball because the numbers do not start at 0.

LESSON 9-6 Puzzles, Twisters & Teasers
Truth or Consequences!

Use the two graphs to answer the questions. Match the letters to your answers to solve the riddle.

1. What is the scale on the magazine circulation graph? __2000__ S

2. What is the scale on the company salaries graph? __10,000__ P

3. The circulation of Great Gardens is how many times greater than that of National Life? __2__ M

4. The owner's salary appears to be how many times greater than the secretary's? __5__ U

5. The circulation of Cycling is how many thousands more than ET? __6__ P

What shoes should you wear when your basement is flooded?

P	U	M	P	S
10,000	5	2	6	2000

LESSON 9-7 Practice A
Scatter Plots

1. Use the given data to make a scatter plot.

Calories and Fat Per Portion of Meat & Fish

	Fat (grams)	Calories
Fish sticks (breaded)	3	50
Shrimp (fried)	9	190
Tuna (canned in oil)	7	170
Ground beef (broiled)	10	185
Roast beef (relatively lean)	7	165
Ham (light cure, lean and fat)	19	245

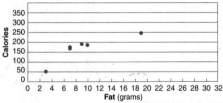

Do the data sets have a positive, a negative, or no correlation?

2. The size of the bag of popcorn and the price of the popcorn

   positive correlation

3. The increase in temperature and number of snowboards sold

   negative correlation

4. Use the data to predict how much money Tyler would be paid for babysitting $7\frac{1}{2}$ hours.

Amount Tyler Earns Babysitting

Hours	1	2	3	4	5	6	7	8
Amount	$4	$8	$12	$16	$20	$24	$28	$32

According to the data, Tyler would get paid $ __30__ for babysitting $7\frac{1}{2}$ hours.

LESSON 9-7 Practice B
Scatter Plots

1. Use the given data to make a scatter plot.

Tall Buildings in U.S. Cities

Building	City	Stories	Height (meters)
Sears Tower	Chicago	110	442
Empire State Building	New York	102	381
Bank of America Plaza	Atlanta	55	312
Library Tower	Los Angeles	75	310
Key Tower	Cleveland	57	290
Columbia Seafirst Center	Seattle	76	287
NationsBank Plaza	Dallas	72	281
NationsBank Corporate Center	Charlotte	60	265

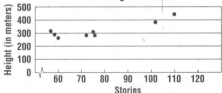

Do the data sets have a positive, a negative, or no correlation?

2. The temperature outside and the number of ice cream cones sold

   positive correlation

3. The amount of time spent in the bathtub and the temperature of the bath water

   negative correlation

4. Use the data to predict the percent of Americans owning a home in 1955.

Percent of Americans Owning Homes

Year	1950	1960	1970	1980	1990
Percent	55.0%	61.9%	62.9%	64.4%	64.2%

According to the data, about __58.5__ % of Americans owned a home in 1955.

Practice C
9-7 Scatter Plots

1. Use the given data to make a scatter plot of length and maximum depth of the largest lakes of North America.

Largest Lakes of North America

Lake	Area (square miles)	Length (miles)	Maximum Depth (ft)
Superior	31,700	350	1330
Huron	23,000	206	750
Michigan	22,300	307	923
Great Slave	11,031	298	2015
Erie	9910	241	210
Ontario	7340	193	802
Athabasca	3064	208	407
Reindeer	2568	143	720

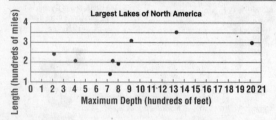

Do the data sets have a positive, a negative, or no correlation?

2. The amount of time spent exercising and the number of calories burned

   positive correlation

3. The discount percent of an item and the total cost of the item

   negative correlation

4. Use the data to estimate the median four-person family income in 1982.

Median Four-Person Family Income

Year	1975	1980	1985	1990	1995	1999
Income	$15,848	$24,332	$32,777	$41,151	$49,687	$59,981

According to the data, the median four-person family income was about **$28,000** in 1982.

Reteach
9-7 Scatter Plots

Two sets of data can be graphed as points in a **scatter plot**. If there is a relationship between the data sets, a **line of best fit** can be drawn.

Positive correlation: both sets of data increase together.

Negative correlation: values of one set increase while values of the other set decrease.

No correlation: points neither increase nor decrease together.

Make a scatter plot. Include a line of best fit if there is a correlation. Describe the correlation.

1.

Time (hours)	1	2	2.5	6
Distance (miles)	50	150	175	270

The values for time are: increasing
distance are: **increasing**
So, there is a **positive** correlation.

2.

Number of Workers	6	4	2	1
Number of Days	1	2	5	7

The values for the number of workers are: **decreasing**
the number of days are: **increasing**
So, there is a **negative** correlation.

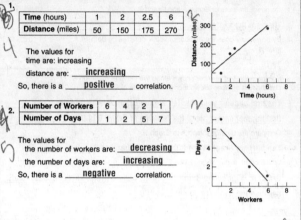

Challenge
9-7 This Fits Nicely!

When two sets of data show a correlation, you can draw a **line of best fit** that approximates a trend.

Here are some data relating the gestation periods of selected animals to their average life spans. The data are separated into 3 equal sets.

	Set I			Set II			Set III	
Gestation (days)	31	61	68	105	151	167	285	330	365
Longevity (years)	13	12	4	5	8	20	15	20	12

1. Determine the **median-median point** for each set of points by getting the median value for the gestation values and the median value for the longevity values.

The median-median point for
Set I is: **(61, 12)** Set II is: **(151, 8)** Set III is: **(330, 15)**

2. Make a scatter plot for the given data. Describe the correlation. **positive**

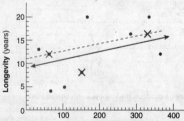

3. Using an X for each, plot the three median-median points on your graph.

Using a ruler, draw a dotted line through the median-median points for Sets I and III. **See graph**

Keeping the ruler at the level of the dotted line, estimate the vertical distance between the dotted line and the median-median point for Set II. Then slide the ruler down about one-third this distance. Draw a solid line parallel to the dotted line. This solid line is called the median-median line and it is a line of best fit for the given data. **See graph**

Problem Solving
9-7 Scatter Plots

Use the data given at the right.

1. Make a scatter plot of the data.

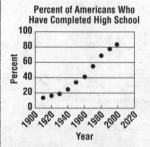

Percent of Americans Who Have Completed High School

Year	Percent
1910	13.5
1920	16.4
1930	19.1
1940	24.5
1950	34.3
1960	41.1
1970	55.2
1980	68.6
1990	77.6
1999	83.4

2. Does the data show a positive, negative or no correlation?

   Positive

3. Use the scatter plot to predict the percent of Americans who will complete high school in 2010.

   About 90%

Choose the letter for the best answer.

4. Which data sets have a positive correlation?
   A The length of the lines at amusement park rides and the number of rides you can ride in a day
   B The temperature on a summer day and the number of visitors at a swimming pool
   C The square miles of a state and the population of the state in the 2000 census
   D The length of time spent studying and doing homework and the length of time spent doing other activities

5. Which data sets have a negative correlation?
   F The number of visitors at an amusement park and the length of the lines for the rides
   G The amount of speed over the speed limit when you get a speeding ticket and the amount of the fine for speeding
   H The temperature and the number of people wearing coats
   J The distance you live from school and the amount of time it takes to get to school

Reading Strategies 9-7
Vocabulary Development

A **scatter plot** shows whether two sets of data are related. The data are shown as points.
This scatter plot shows data about a child's age and his or her shoe size.
By looking at both, you can see if there is a relationship between the two sets of data.

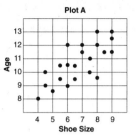

Plot A

Answer each question.

1. What information is shown along the left side of the plot?
   age

2. What is the age range of children shown on this plot?
   between 8 and 13   or 5 years

3. What information is listed along the bottom of the plot?
   shoe size

4. What is the range of shoe sizes shown on this plot?
   between 4 and 9   or 5 shoe sizes

5. Describe the pattern of points on the plot.
   Possible answer: The points move in an upward direction.

6. As a child's age increases, what happens to his or her shoe size?
   It increases also.

7. Is there a relationship between a child's age and his or her shoe size?
   yes

Puzzles, Twisters & Teasers 9-7
The Plot Thickens!

Identify each pair of data sets as having a positive correlation, negative correlation or no correlation. Use the letters of your answers to solve the riddle.

1. The number of pages you have read in a book and the number of pages remaining
   T Positive Correlation   **Ⓞ** Negative Correlation   D No Correlation

2. The day of the month and the wind speed
   T Positive Correlation   O Negative Correlation   **Ⓓ** No Correlation

3. The age of a car and its selling price
   O Positive Correlation   **Ⓣ** Negative Correlation   D No Correlation

4. The weight of a vehicle and its gas mileage
   L Positive Correlation   **Ⓜ** Negative Correlation   P No Correlation

5. The outside temperature and the number of people in attendance at the beach
   Ⓨ Positive Correlation   L Negative Correlation   S No Correlation

6. The month of the year and the number of birthdays in a certain month
   S Positive Correlation   O Negative Correlation   **Ⓦ** No Correlation

7. The population of a state and the number of senators
   L Positive Correlation   R Negative Correlation   **Ⓐ** No Correlation

8. The length of your hair and the number of days since your last haircut
   Ⓝ Positive Correlation   T Negative Correlation   K No Correlation

9. The number of hours spent studying and the test score received
   Ⓡ Positive Correlation   S Negative Correlation   B No Correlation

Which two days of the week start with T?

T O D A Y   A N D
3 1 2 7 5   7 8 2

T O M O R R O W
3 1 4 1 9 9 1 6

Practice A 9-8
Choosing the Best Representation of Data

1. Which graph is a better display of the number of vacation days certain countries have in a year?

 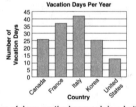

   The data compares different numbers of days, so the bar graph is a better display.

2. Which graph is a better display of the distribution of cats' weights?

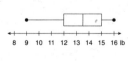 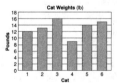

   The question asks about the distribution of data, so the box-and-whisker plot is the better display.

3. A scientist measured the lengths of 10 earthworms. The table shows her data. Choose an appropriate data display and draw the graph.

Earthworm Lengths (cm)				
8	14	9	10	9
11	7	10	12	9

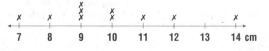

Practice B 9-8
Choosing the Best Representation of Data

1. Which graph is a better display of the number of students in a class who chose math as their favorite subject?

   The question asks to find a share, or part of a whole, so the circle graph is a better display.

2. Which graph is a better display of the change in the number of cell telephone subscribers?

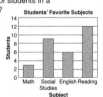

   The data show changes over time, so the line graph is the better display.

3. The table shows the heights of players on a school basketball team. Choose an appropriate data display and draw the graph.

Heights of Basketball Players (in.)			
70	64	68	71
61	68	65	73

Possible answer:

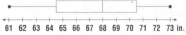

LESSON 9-8 Practice C
Choosing the Best Representation of Data

1. Which graph is a better display of the frequency of students' scores on a math quiz?

The question asks about frequencies, so the histogram is a better display.

2. Which graph is a better display of the U.S. population after the last five censuses?

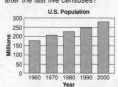 

The data show changes over time, so the line graph is probably the better display.

3. The table shows the ages and heights of corn plants. Choose an appropriate data display and draw the graph.

Age (days)	4	6	3	8	6	9	4	5	7	9
Height (cm)	3	4	3	7	5	8	4	7	4	6

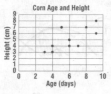

LESSON 9-8 Reteach
Choosing the Best Representation of Data

A line graph is a good way to show changes over time. The amount of money in Tasha's account changes from month to month.

A bar graph is a good way to compare different quantities. The amounts of money in the students' bank accounts are being compared.

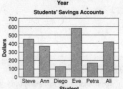

A circle graph is a good way to compare parts of a whole. Lionel's assets are the whole; the parts are his savings account, checking account, and mutual funds.

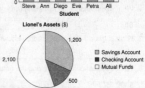

Determine which kind of graph would be best to display the described data. Write *line graph*, *bar graph*, or *circle graph*.

1. Total number of hits by each member of a softball team

   bar graph

2. Number of miles that a boy rides his bike each week

   line graph

3. Daily profits of a hot dog stand at a county fair

   line graph

4. Part of a hot dog stand's profits that come from popcorn sales

   circle graph

LESSON 9-8 Challenge
What's Going On?

Invent a situation for the graph. Label the parts of the graph and describe the data it displays.

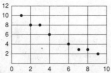

Labels and answers will vary, but should describe two sets of data with a negative correlation.

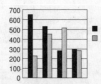

Labels and answers will vary, but should describe two series of data being compared.

Labels and answers will vary, but should describe data set with four parts.

LESSON 9-8 Problem Solving
Choosing the Best Representation of Data

Write what kind of graph would be best to display the described data.

1. Numbers of times that members of track team ran a mile in the following intervals: 4 min 31 s to 4 min 40 s, 4 min 41 s to 4 min 50 s, 4 min 51 s to 5 min, 5 min 1 s to 5 min 10 s

   histogram

2. Distribution and range of students' scores on a history exam

   box-and-whisker

3. Relationship between the amounts of time a student spent on her math homework and the numbers of homework problems she solved

   scatter plot

4. Total numbers of victories of eight teams in an intramural volleyball league

   bar graph

5. Part of calories in a meal that come from protein

   circle graph

6. Numbers of books that a student reads each month over a year

   line graph

Choose the letter for the best answer.

7. A bar graph is a good way to display
   A data that changes over time.
   B parts of a whole.
   C distribution of data.
   D comparison of different groups of data.

8. A circle graph is a good way to display
   F range and distribution of data.
   G distribution of data.
   H parts of a whole.
   J changes in data over time.

9. A scatter plot is a good way to display
   A comparison of different groups of data.
   B distribution and range of data.
   C the relationship between two sets of data.
   D parts of a whole.

10. A box-and-whisker plot is a good way to display
    F range and distribution of data.
    G the relationship between two sets of data.
    H data that changes over time.
    J parts of a whole.

LESSON 9-8 Reading Strategies
Understand Vocabulary

Refer to the graph at right.

1. What does the title of the graph tell about the graph?

   That the graph shows the record of a boy's soccer team.

2. What does the legend tell about the graph?

   which bars show wins and which bars show losses

3. What does the scale of the graph show?

   number of wins and losses by the soccer team

4. What is the range of the scale?

   _5_

5. Which month did the team have the most losses?

   July

6. Which month did the team have the most wins?

   October

LESSON 9-8 Puzzles, Twisters & Teasers
The Best Displays

Across

1. The kind of display preferred by cats
2. When you want to see the distribution
3. How a weightlifter might keep track of his progress

Down

1. Is there a relationship between the data sets?
2. Time changes everything
3. When you want to see who ate the biggest part of the pizza
4. What is the frequency?

Across: 1. BOXANDWHISKER 2. LINEPLOT 3. BARGRAPH

Down: 1. SCATTERPLOT 2. LINEGRAPH 3. CIRCLEGRAPH 4. HISTOGRAM